Diffractive Optics
Design, Fabrication, and Test

Donald C. O'Shea

Thomas J. Suleski

Alan D. Kathman

Dennis W. Prather

Tutorial Texts in Optical Engineering
Volume TT62

SPIE PRESS

A Publication of SPIE—The International Society for Optical Engineering
Bellingham, Washington USA

Library of Congress Cataloging-in-Publication Data

Diffractive optics : design, fabrication, and test / Donald C. O'Shea...[et al.]
    p. cm. — (SPIE tutorial texts ; v. TT62)
Includes bibliographical references and index.
ISBN 0-8194-5171-1 (soft cover)
1. Optical instruments–Design and construction. 2. Lenses–Design and construction. 3.
Diffraction. I. O'Shea, Donald C. II. Series.

TS513.D54 2003
621.36–dc22   2003055704
                                                                                  CIP

Published by

SPIE—The International Society for Optical Engineering
P.O. Box 10
Bellingham, Washington 98227-0010 USA
Phone: 360.676.3290
Fax: 360.647.1445
Email: spie@spie.org
www.spie.org

Contents

Preface

This work is based on a series of short courses in diffractive optics that have been presented at Georgia Institute of Technology since 1994. The course was started as a hands-on workshop that provided basic theory on diffractive optics and then allowed participants to progress through a series of exercises on the design, fabrication, and testing of diffractive optical elements (DOEs). This type of course was difficult to present because of the intensive support required for the labs. When one of the authors (TJS) and two of his fellow graduate students got their doctorates, we lost all our good, cheap help and we had to radically change the course. The new offering relied on additional lectures and demonstrations to replace the exercises. When we finished with this revision, we knew that the material in the restructured course could serve as the basis for a text on diffractive optics.

This book is intended to provide the reader with the broad range of materials that were discussed in the course. We assume the reader is familiar with basic computational techniques and can stand the sight of an integral or two. It is not our intention to overwhelm the reader with long derivations or provide detailed methods for specific engineering calculations. Instead we introduce the concepts needed to understand the field. Then a number of simple examples, which someone can use as a check on their initial baseline calculations, are presented. While this text is not a "cookbook" for producing DOEs, it should provide readers with sufficient information to be able to assess whether the application of this technology would be beneficial to their work and give them an understanding of what would be needed to make a DOE.

In the work presented in the course we describe two methods of generating the binary masks needed to produce the diffractive optics elements. One is a costly technique that yields state-of-the-art results and is the basis for most commercial production. The second, exploited by the diffractive optics group at Georgia Tech, uses standard desktop publishing techniques and PostScript output to produce masks with modest feature sizes. The latter technique is useful for simple prototyping and for educational demonstrations. In this text we have separated the two approaches by discussing the high-resolution technique as the primary mask fabrication path. For those who want to get their feet wet, we have provided a few boxes set off from the main narrative that describe how the PostScript methods can replace the standard techniques at a savings of time and money, but with a loss of performance.

After a brief introductory chapter on the field, we provide a description of the theoretical basis for the operation of diffractive optical devices. In most cases a scalar theory description will suffice, particularly as an introduction. However, as the wavelength of the radiation approaches the size of the various features in the element, a more precise theory that includes a vector description of the electric fields in the vicinity of the surface is required. Next, a series of chapters describe the procedures used to design elements that can be incorporated into conventional lens designs, in addition to procedures for designing periodic structures and unconventional devices. This is followed with a description of the various steps in the fabrication and test of diffractive optical elements. Finally, we provide a short survey of a number of applications in which these devices are making an impact on today's technology.

We would like to acknowledge the contributions to the course made by some of the earlier lecturers and assistants. Tom Gaylord at Georgia Tech and Joe Mait of the Army Research Laboratory provided lectures in scalar and vector theory. Willie Rockward and Menelous Poutous (along with TJS) helped put together the exercises for the workshop and conducted the labs. The authors also wish to thank their wives, who put up with a lot. They never have figured out how we could argue so fervently over those little ripples in a piece of quartz.

Donald C. O'Shea   June 2003
Thomas J. Suleski
Alan D. Kathman
Dennis W. Prather

Diffractive Optics

Design, Fabrication, and Test

Chapter 1

Introduction

1.1 Where Do Diffractive Elements Fit in Optics?

When optical engineers talk about controlling light through diffraction, a number of terms are used: binary optics, kinoforms, computer-generated holograms, holographic optical elements, and plain old diffraction gratings. To begin this description of diffractive elements and their applications, we need to sort out the differences between these terms and settle on what we will call them in this text. For the purposes of this discussion, the optical components that are the subject of this text will be called diffractive optical elements (DOEs).

Also, it would be useful to know how these terms relate to one another. One way of organizing all of them is shown in a diagram in Fig. 1.1. Overall, these concepts can be arranged within the field of diffractive optics. This is the field, distinguished from physical optics, that describes the control and generation of wavefronts by segmenting initial wavefronts and redirecting the segments through

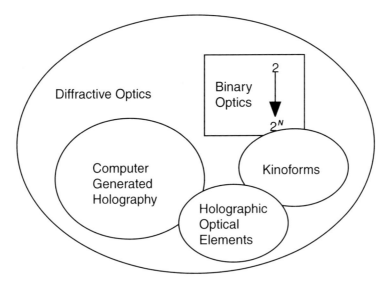

Figure 1.1 Schematic organization of the various types of diffractive optical elements within the field of diffractive optics.

the use of interference and phase control. As such, diffractive optics is a subfield of physical optics, which deals with all forms of propagation of electromagnetic waves.

Some of the different types of elements in the diagram are the following:

- **Diffractive Optical Element:** This is a component that modifies wavefronts by segmenting and redirecting the segments through the use of interference and phase control.
- **Kinoform:** This is a DOE whose phase-controlling surfaces are smoothly varying.
- **Binary Optic:** A DOE with a discrete number of phase-controlling surfaces. The simplest form is produced using one lithographic mask and has only two surfaces, which introduce either a 0 or π-phase difference on the incident wavefront. When N masks are used, a multilevel binary optic can be generated, usually resulting in 2^N phase levels.
- **Computer-Generated Hologram:** This is a DOE generated by reducing a calculated interference pattern to a series of phase or amplitude masks. It is similar to other diffractive optics except that the useful wavefronts tend to be one of several orders generated by the pattern.
- **Holographic Optical Element:** This is a DOE generated by the interference of two wavefronts to produce a component that will be used to act as an optical component. (The only distinction between this element and a hologram is that a hologram is usually intended to record some elaborate scene.)

Before presenting a short survey of different types of DOEs, some concepts must be introduced to assist in understanding how diffractive elements work. The propagation of light through some optical components can be most easily understood as bundles of rays that are transmitted through an optical system to form images or to produce light patterns. However, to understand DOEs, it is best to think in terms of wavefronts, the continuous surfaces perpendicular to the paths of light rays on which the electric field has the same phase and usually the same amplitude. For example, light rays coming from a long distance away (infinity) will be parallel to each other and the corresponding wavefronts perpendicular to the light rays. When these plane waves are focused by a positive lens, they are converted into converging spherical waves centered on the focal point.

1.2 A Brief Survey of Diffractive Optics

Diffractive optical elements can be used in conjunction with other optical components or they can produce effects on their own depending on what the optical engineer is trying to achieve for a particular design goal. One way of surveying the range of diffractive optical elements is to place them on a continuum between a classic optical element (a lens) and a DOE with no classic counterpart—a general wavefront transformer. This continuum is displayed in Fig. 1.2.

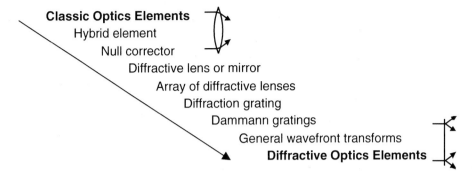

Figure 1.2 Range of elements that contain diffractive surfaces.

At the classic end of this continuum is the hybrid lens, consisting of a conventional refractive lens with a diffractive structure etched into one of its surfaces (Fig. 1.3). As will be shown in Chapters 4 and 9, the diffractive surface can be used to correct image aberrations and color aberrations in a lens in a manner similar to the use of aspheric surfaces and additional refractive components. For example, the dispersion of a diffractive surface, or the variation of optical power with wavelength, is so great that a weak diffractive lens can be added to a standard lens to provide the required color correction. More generally, the flexibility of designing and making diffractive surfaces adds a useful tool to a lens designer's toolbox.

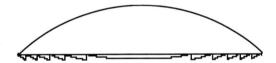

Figure 1.3 Hybrid lens consisting of both refractive and diffractive surfaces.

This flexibility in constructing wavefronts also gives an optical designer the ability to use DOEs as optical elements all by themselves. Fabricated on flat surfaces, the microscopic structures can act as converging [Fig. 1.4(a)] and diverging lenses and mirrors. In situations where size or weight are critical, diffractive lenses bring substantial advantages. DOEs can also be used to correct aberrated wavefronts generated by some optical systems [Fig. 1.4(b)]. If an aberrated wavefront

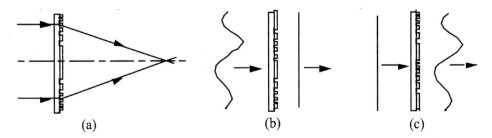

Figure 1.4 Diffractive optical elements. (a) Diffractive lens. (b) Wavefront corrector. (c) Null corrector for optical testing.

can be characterized, it is probably possible to create a diffractive element to flatten that wavefront. Conversely, they can also act as null correctors in optical testing, providing the precise wavefront that, when transmitted through the optical system under test, should generate a simple, easily detected plane or converging wavefront [see Fig. 1.4(c)] so that any deviations can provide information on residual errors in the tested optics.

Two other DOEs that have counterparts in conventional optics are shown in Fig. 1.5. One is a beam sampler, a device that diverts a small fraction of a wavefront from the direction of propagation for testing and control while permitting most of the beam to be transmitted without modification [Fig. 1.5(a)]. The other is an array of optical components, such as a lens array [Fig. 1.5(b)].

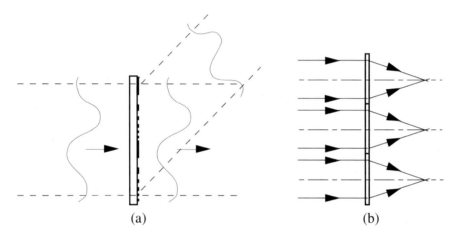

(a)   (b)

Figure 1.5 Diffractive optical elements. (a) Beam sampler. (b) Diffractive lens array.

Another class of DOEs, called *pattern generators*, is based upon the diffraction grating. Most people tend to think of a diffraction grating as a device for separating light into its spectrum of colors. If the incident light is monochromatic, however, the periodic structure produces an array of regularly spaced beams. Their direction is given by the grating equation,

$$m\lambda = \Lambda \sin \theta_m, \tag{1.1}$$

where λ is the wavelength of light, Λ is the grating period, and θ_m is the angle of the diffracted order m. If the grating is illuminated by white light, it separates the light into its spectrum of colors, the angles again given by Eq. (1.1), but where the value of λ is not constant. In contrast to conventional gratings, the lithographic technology used to generate these DOEs can be employed to control the relative intensities of the individual beams by shaping the profile within each grating period.

One particularly useful application of pattern generators is the optical interconnect. By engineering the diffraction efficiency of a diffractive optical element in its various orders, light from a number of sources, such as laser diodes, can be directed to a number of detectors, as shown in Fig. 1.6. Among other things, these

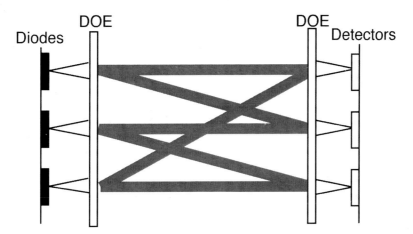

Figure 1.6 A simple optical interconnect made of two diffractive optical elements can provide multiple paths between arrays of diodes and detectors.

interconnects can be used for switching applications and as the basis for arithmetic elements in an optical computer.

As fabrication technologies advance, additional applications of diffractive optics emerge. If we construct a grating with smaller and smaller grating periods, the beam fanout angles increase. Eventually there is a value of a that is smaller than the wavelength of the illuminating radiation. Below that, according to Eq. (1.1), there are no diffracted orders. It would seem that such a device would be useless, but as will be shown, these subwavelength structures can be used to generate antireflective (AR) surfaces and birefringent elements.

For example, consider the structure shown in Fig. 1.7. It is fabricated directly as part of the substrate of some optical element. If the shapes of the triangular structures are made properly, the serrated surface will transmit light of a particular wavelength with no reflections. It behaves in a manner similar to that of sound-absorbing "egg crate" panels or anechoic walls that do not reflect radar signals. This patterned surface acts much like conventional antireflection films, but with a number of advantages over standard multilayer coatings. There are no thermal expansion mismatches or surface adhesion problems. It has a large field of view and bandwidth and is lightweight and compact compared with other filters. Finally, it is fabricated from isotropic materials and can accommodate substantial refractive index mismatches. Because the profile of this surface is similar to that found in

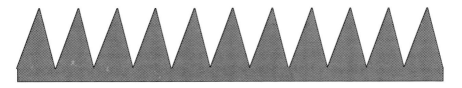

Figure 1.7 Moth's-eye structure, a diffractive surface with a repeat distance less than the wavelength of light. If it is made with the proper profile, this surface acts as an antireflecting surface.

the eyes of moths, it is often called a *moth's-eye structure*. Other subwavelength structures can be designed to function as filters and polarizers.

At the end of the continuum shown in Fig. 1.2 are general wavefront transforms. A DOE can be used to map an incoming wavefront into one or more resulting wavefronts, with sometimes startling results. Lens functions can be combined with fanout gratings to map a collimated input beam into an array of spots (Fig. 1.8). Several functions can be multiplexed together to create bifocal lenses as well. Complex fanout gratings have even been used to create DOE portraits. Some of the most sophisticated general wavefront transforms are used for optical computing. Diffractive optics are used to perform coordinate transformations and solve pattern recognition problems at, literally, the speed of light.

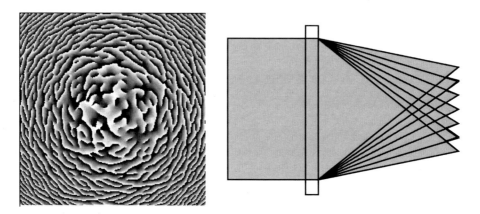

Figure 1.8 A 16-level fanout-grating function combined with a lens function will map a collimated beam to multiple spots.

1.3 A Classic Optical Element: The Fresnel Lens

The imaging of any point on an object can be represented either by the refocusing of rays diverging from the point onto the image or by reconverging the spherical wavefronts diverging from the object point, as shown in Fig. 1.9.

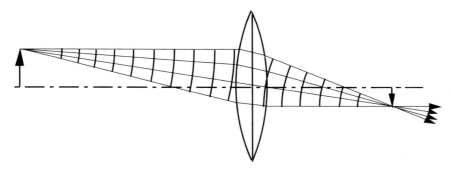

Figure 1.9 Imaging with a lens can be represented as either a redirection of bundles of rays or as a converging of spherical wavefronts diverging from an object point.

As a way of understanding the differences between classic optical elements and diffractive elements, the operation of a Fresnel lens is compared with a diffractive lens. A plano-convex lens and its Fresnel lens counterpart, both classical elements, are shown in Fig. 1.10. A bundle of parallel rays incident upon the lens will be focused by the plano-convex lens to its focal point. The Fresnel lens will do the same. In effect, the Fresnel lens can be considered as being created by removing slabs of glass that do not contribute to the bending of the light rays to a focal point. The surface profile of the lens that is responsible for the optical power of the element is preserved, but the volume and weight of the lens are reduced.

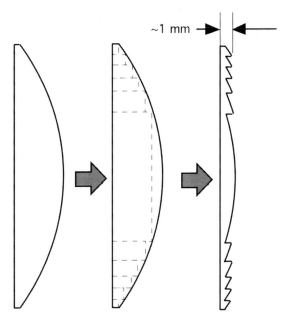

Figure 1.10 Generation of a Fresnel lens from a plano-convex lens. The surface profile of the lens is preserved while the volume of the lens is reduced.

Imaging occurs as with a regular lens in that a bundle of rays from one point on an object is imaged to a corresponding point in the image plane. It is possible to understand the process as just a set of ray traces through two surfaces, with the application of the law of refraction at each surface.

1.4 Light Treated as a Propagating Wave

When light passes through a narrow slit, it cannot be modeled using rays. Instead, it is necessary to determine how a wavefront will propagate as it passes through the slit. A Dutch scientist, Christian Huygens, showed that wave propagation could be understood using a simple graphical construction. He employed the point source as the basis for his technique. A point source is an idealization of a small source that emits radiation in all directions. Its output is a series of expanding spherical wavefronts centered about the point source. These wavefronts expand at the speed of light in ever-larger spherical shells.

Huygens construction is a three-step procedure that determines the shape of a wavefront as it propagates through space. It is illustrated in Fig. 1.11.

1. The original wavefront is evenly populated by a series of point sources [Fig. 1.11(a)].
2. The point sources all emit spherical wavefronts, all in phase with one another [Fig. 1.11(b)].
3. The envelope of the wavefronts after some time t gives the shape of the propagated wavefront after the same time [Fig. 1.11(c) and 1.11(d)].

One of the consequences of Huygens construction, as can be seen in Fig. 1.11(d), is that the propagating wave extends beyond the width of the aperture. This is the consequence of diffraction—any obstruction causes light to spread. For example, a circular aperture provides an interesting case. As will be shown, the diameter of the circle can profoundly affect the strength and distribution of the light behind the aperture.

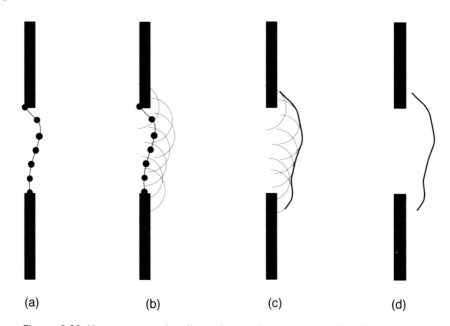

(a)   (b)   (c)   (d)

Figure 1.11 Huygens construction of wavefront propagation through a slit.

If a screen is located at a point P at a distance r_0 behind a circular aperture and on a perpendicular line through its center (Fig. 1.12), there will be a series of circles in the aperture where the distances from the aperture to point P are integer half-wavelengths more than r_0. The central circle and these surrounding annuli form a set of areas called *Fresnel zones*.

It can be shown that each of these zones contains exactly the same area, so they all contribute equally to the amount of light falling on the screen at P. The effect of each of the zones on the amount of light falling on the screen on-axis at r_0 can be

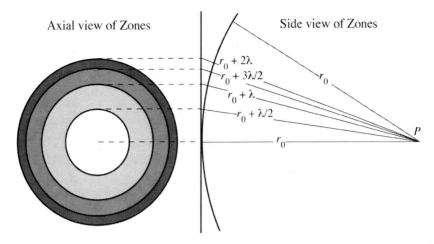

Figure 1.12 Fresnel zones.

gauged by another graphic technique that employs "phasors." Each of these half-wavelength zones can be divided into a series of subzones of equal areas. For our illustration, a zone is subdivided into six subzones. Because the average path length for each of the subzones differs by the same phase angle ($\pi/6$), the combined effect of the subzones can be plotted as a series of vectors of equal length with the phase angle between the two neighboring subzones (Fig. 1.13). The vector sum of the first set of subzones adds up to the contribution of the first zone. The second zone generates the same contribution as the first zone shown in the figure. However, each subzone in the first zone has a corresponding subzone in the second zone, so that if the aperture consisted of only these two zones, they would cancel each other out!

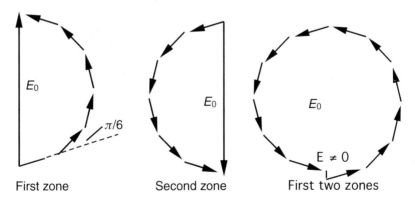

Figure 1.13 Phasor addition of subzones to determine the effect of the first two Fresnel zones of an aperture at some distance from a screen.

As more and more zones are added, the amplitude of the on-axis electric field at the screen will oscillate between minima near zero and maxima nearly twice the value of the electric field that would fall on the screen if there were no aperture between the source and the screen. The contributions for the zones further out are not as great, so as the aperture is opened, the sum of the phasors becomes a spiral of decreasing radius that ends in the center at E_0, the value of an unobstructed field.

But what if every other zone were blocked, so that the canceling contributions were eliminated? This situation is illustrated in Fig. 1.14. Light is transmitted through the first (central), third, and fifth zones, but is blocked for the second, fourth, sixth, and all higher-order zones [Fig. 1.14(a)]. Therefore the contributions from the three zones add in phase [Fig. 1.14(b)], so that the resulting electric field is $6E_0$ and the on-axis irradiance at P is 36 times that for the unobstructed field. The pattern that caused this increase of light intensity is called a *Fresnel zone plate* and it acts as a diffractive lens.

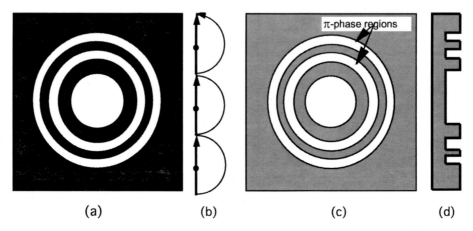

Figure 1.14 Fresnel plates. (a) Fresnel zone plate. (b) Phasor contributions from the three odd-numbered zones. (c) Fresnel phase plate. (d) Profile through the center of the Fresnel phase plate.

If instead of blocking the alternate zones, the phase of these zones is shifted by π (180 deg), then the light that had been blocked by these zones will now be added to the contributions from the odd-numbered zones. This strategy allows more of the energy incident upon the element to be directed to P. The contributions from the three odd-numbered zones and the three shifted zones all add in phase, so the resulting electric field is $12E_0$, and the on-axis irradiance at P is 144 times that for the unobstructed field. This device is called a *Fresnel phase plate*. It will be shown that the depth of the π-phase zones is $\lambda/2(n-1)$, which introduces the π-phase shift.

A Fresnel phase plate is a diffractive optical element that focuses light, but it is not very efficient. As we will show later, only about 41% of the transmitted light would be focused onto the axis of this "lens." It also acts as a negative lens, so another 41% of the light diverges from the element. When additional phase levels are added to this two-level structure, more of the transmitted energy is focused on P. In Fig. 1.15, several multilevel profiles for this diffractive lens are shown, along with the limiting case of a smooth profile, the kinoform.

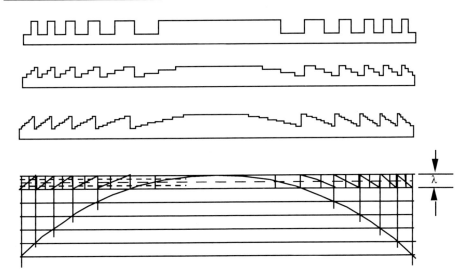

Figure 1.15 Profiles of a diffractive lens ranging from a two-level Fresnel phase plate (top) and four-level and eight-level structures (center) to a kinoform profile. It shows the fracturing of the profile to generate the transitions for the multilevel structures.

1.5 A Physical Optics Element: The Blazed Grating

Consider a prism that refracts a beam through an angle. In Fig. 1.16(a) the progress of some rays and their corresponding wavefronts through a right-angle prism is shown. If the prism is sliced into one-wavelength-high pieces and if all the rectangular sections that are one wavelength thick (and do not contribute to a change in path length) are removed, the resulting element looks like the element in Fig. 1.16(b). This is known as a blazed grating. It directs all of the light at that particular wavelength into the first order. Both elements deflect light. The prism does this by refraction. Any variation in the direction of light with wavelength in the prism is due to the change of refractive index with wavelength, a property called *dispersion*. In contrast, the blazed grating produces the same effect by diffraction. The dispersion of the grating is determined by its periodicity and is governed by the

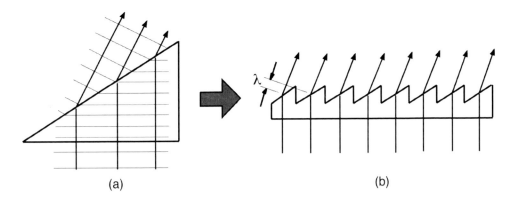

(a)   (b)

Figure 1.16 Deflecting wavefronts. (a) Prism. (b) Blazed grating.

grating equation [Eq. (1.1)]. Figure 1.16 is one of the simplest demonstrations of the difference between a classic optical element and a diffractive optical element.

The concept of a blazed grating can be applied to a diffractive optical element. We can gain additional insight into the operations of a diffractive optical element by observing the results of the plane waves passing through a kinoform lens. In Fig. 1.17, a number of plane waves are incident from the left onto a kinoform with phase changes of 2π at the transitions. As each of these wavefronts traverses the kinoform, it is reformed into a series of one-wavelength-deep arc segments, shown at the right of the kinoform. Each of set of fractured wavefronts, following one behind the other, aligns with the neighboring segments to produce a series of concave spherical wavefronts that focus the light. In effect, this diffractive element "stitches" wavefronts together to produce a new wavefront.

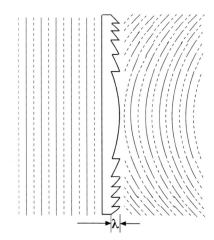

Figure 1.17 Operation of a diffractive lens as combiner of wavefronts.

Another approach to understanding the operation of a diffractive optical element is to regard it as an element made up of variably spaced diffraction gratings that deflect the local segment of the wavefront. Note that the top and bottom edges of the kinoform profile in Fig. 1.17 has the same "sawtooth" profile as the profile of the blazed grating in Fig. 1.16.

In the example given here, the diffractive lens transforms plane waves into spherical waves. The focal length of the lens depends only on the profile. In fact, almost any element profile can be fabricated, as we will describe later on. One way of thinking about this is to consider the wavefront on the right side of Fig. 1.18. It has no particular symmetry or shape. It could be produced from a material with a profile to the left of the wavefront. That profile can be sliced into 2π-phase segments, as shown to the left of it. On the far left is the element that will produce the wavefront on the right. Therefore it is possible to generate almost any wavefront. In a sense, using diffractive optics, wavefronts can be made to order. There are some limits to this assertion, which have to do with the size of the features that can be made and their size in relation to the wavelength of light. They are discussed in more detail in later chapters.

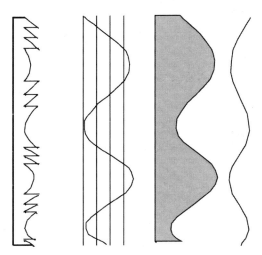

Figure 1.18 A "made-to-order" wavefront. Starting from the right, the wavefront to be synthesized from a plane wave requires the profile on its left. This profile is then sliced into wavelength sections and then the sections are reduced to the one-wavelength-high diffractive profile on the left.

1.6 Fanout Gratings

In addition to making efficient gratings through blazing, it is possible to generate gratings that can modify wavefronts in ways that were not possible before. Consider the problem of splitting a laser beam into a certain number of beams. It is possible to do this conventionally using a combination of beamsplitters, but the method is very awkward. The number of mirrored surfaces increases with the number of beams, and the placement of the mirrors becomes difficult. In comparison, a diffractive optical element can produce a number of beams of equal intensity in a number of orders. These are referred to as fanout gratings. In its simplest form a fanout grating is a two-level phase grating called a *Dammann grating*. A Dammann grating has a number of phase transitions between the two levels in each grating period, Λ, as illustrated in Fig. 1.19. In this example, the grating generates five beams of equal intensity plus some other beams of lesser intensity.

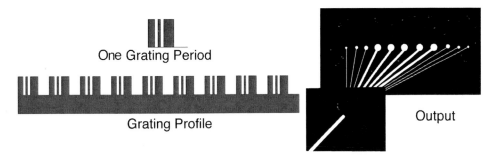

Figure 1.19 A Dammann grating that splits a beam into five beams of equal intensity.

1.6.1 Designing a fanout grating

Fanout grating designs can range from simple to very sophisticated. The simplest type of Dammann grating is produced with a single etch (two-level; usually 0 and π phases), and is designed so that the zero order and the first N orders all have equal intensity. For binary grating structures, the intensity of the orders on both sides of the zero order is equal. Therefore we are designing a Dammann grating that will split a beam into $2N + 1$ beams of equal intensity. It is also possible to generate Dammann gratings with an even number of beams.

If the grating is not restricted to two phase levels, the diffraction efficiencies can be higher (Fig. 1.20). In all but a few cases, multilevel grating structures require some optimization algorithm. The results can be more sophisticated than Dammann gratings. There will be additional discussion on the design of such fanout devices in Chapters 5 and 10.

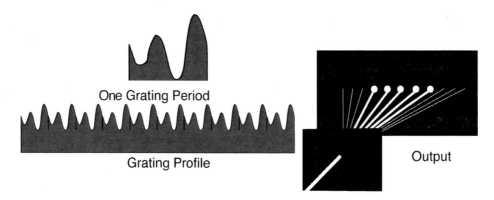

Figure 1.20 High-efficiency grating. The efficiency of Dammann gratings can be increased by using multilevel or smooth profiles.

1.7 Constructing the Profile: Optical Lithography

The key to making a diffractive optical element is the fabrication of a surface profile that will shape an incoming wavefront into the required segments (Fig. 1.17). A number of techniques are explored in Chapters 6 through 8 on fabrication. As an introduction to the technology, we describe one method that generates the desired profile through a series of etches in a substrate using standard photolithographic (photoresist) techniques.

As an example, consider the process that might be used to generate the blazed grating shown in Fig. 1.16(b). First, a binary mask of alternating transparent and opaque bars is fabricated using some type of pattern generator, as shown in the top segment of Fig. 1.21. The mask is laid on a substrate coated with a thin layer of photoresist, which is exposed to ultraviolet light. After the resist is developed, only the unexposed areas of photoresist remain. The substrate is then etched by one of several methods until the optical path difference (OPD) between the etched and unetched levels is one-half of the wavelength of the incident light. The photoresist

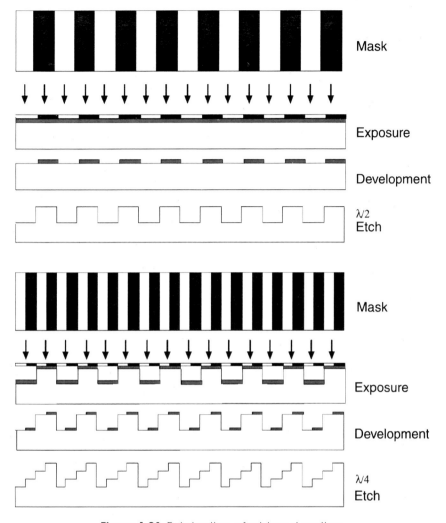

Figure 1.21 Fabrication of a blazed grating.

pattern is then removed and the etched substrate is recoated and reexposed to a second binary mask whose pattern is similar to the first mask, but whose period is half that of the first mask, as shown in the bottom part of Fig. 1.21. After development, the substrate is again etched. This time the depth is one that results in a one-quarter wavelength optical path difference. The result is a four-step sawtoothed profile. Obviously one could continue this process with finer gratings and smaller etching depths to approximate the smooth sawtooth of a blazed grating.

1.8 A Theme

All diffractive optics is in the geometry—the edges of the phase transitions and the heights of the profiles. Materials enter only in determining the step height of the profile and as part of any classic optical design that incorporates a diffractive surface. *Therefore the generation, control, and measurement of the surface profiles are the keys to the design and fabrication of diffractive optics.*

This book is organized so that each aspect of diffractive optics is examined in the sequence used to create an element. After a description in Chapters 2 and 3 of the scalar and vector diffraction theories upon which these devices are based, a number of approaches to the design of various structures are presented in Chapters 4 and 5. Many techniques for rendering the design profiles on a surface are described in Chapters 6 and 7, and methods for measuring the results are given in Chapter 8. In the concluding chapters, a number of applications, while not intended to cover all possible uses, are presented to provide an appreciation of the power and scope of this fascinating new technology.

Chapter 2

Scalar Diffraction Theory

Light is part of the electromagnetic spectrum and therefore is governed by Maxwell's equations. In this chapter we introduce these equations to describe the propagation of light. In later chapters we will use them as a basis for the design of a wide range of diffractive optical elements. Although we begin with the vector form of Maxwell's equations, this representation is difficult to use for most calculations, but they can be simplified by a series of assumptions to their scalar forms, where they are most useful. However, when the wavelength of the incident radiation is comparable to or smaller than the size of the diffractive features, it is necessary to retain the vector forms, and the computations are more difficult. They are discussed in Chapter 3. While several modeling techniques are used in this text, it is important to remember that they are all derived from Maxwell's equations.

When designing an optical element using any mathematical model, it is critical to know that the model is accurate for your design. By understanding the assumptions that are made in the derivation of that model, you can assure yourself that the modeling technique is valid for that element. In Sec. 2.1 of this chapter, the assumptions that are made to reduce Maxwell's equations to the scalar representation are outlined. Then in Secs. 2.2 and 2.3, the use of Fourier analysis to model the performance of a diffractive optical element performance is described. In Sec. 2.4, first-order scalar theory is used to calculate the efficiency of a diffractive structure; Sec. 2.5 discusses the extension of scalar theory to better analyze the efficiency of diffractive structures.

2.1 Rayleigh–Sommerfeld Propagation

Scalar diffraction theory is extremely useful for understanding how many diffractive optical elements operate. It provides a set of simple equations that govern the propagation of light between two planes within an optical system. So if we have been given the transmission function of a diffractive optical element, we can calculate the light distribution at any distance beyond the element. If the complex transmission function of the diffractive element can be determined analytically, then an approach that uses a function, the Rayleigh–Sommerfeld (RS) integral, can be used to calculate the output. Conversely, there are many design tasks where the

desired output is known and an inverse equation is used to determine the form of
the diffractive optical element that will deliver this output. The integral is based on
the scalar diffraction theory, which is particularly useful because the propagation
models can be inverted in this way.

To provide adequate background for this, we present a terse derivation of the
scalar wave equations, beginning with Maxwell's equations for electromagnetic
fields:

$$\nabla \times \mathbf{E} = \frac{\rho}{\varepsilon} \qquad \nabla \times \mathbf{E} = -\frac{\partial \mathbf{B}}{\partial t}$$
$$\nabla \times \mathbf{B} = 0 \qquad \nabla \times \mathbf{B} = \mu\sigma\mathbf{E} + \mu\varepsilon\frac{\partial \mathbf{E}}{\partial t}. \tag{2.1}$$

These four somewhat intimidating equations require both explanation and simplification before they are useful for most optical modeling. The electric field $\mathbf{E}$ and
magnetic field $\mathbf{B}$ are related to one another and the charge density ρ by three material constants: the permeability μ, the permittivity ε, and the conductivity σ. To
simplify Maxwell's equations for optical modeling, we immediately make several
assumptions:[1-3]

(1)   For light to propagate, the medium must be uncharged and nonconducting
      ($\rho = 0$, $\sigma = 0$).
(2)   The material is homogeneous and uniform; μ and ε do not vary with position
      in the material or change with time.
(3)   The material is isotropic, so μ and ε do not vary with orientation.
(4)   The material is linear, so μ and ε do not change in the presence of electric or
      magnetic fields.

These assumptions are true for most optical materials. A charged, conducting
material would absorb light. Materials whose properties do vary in time typically
change very slowly relative to the frequency of light, and therefore can be modeled
as if their properties are static. Space and gases, as well as most liquids and glasses,
are isotropic, while many crystals are not. Assumption 2 does not hold within the
microstructure that defines the diffractive element, but we will not be modeling
light propagation within the microstructure. These assumptions greatly simplify the
mathematics, but they do not limit the applicability of these equations for modeling
propagation from a diffractive element to subsequent planes in an optical system.

By incorporating assumptions 1 to 4, taking the second derivative of both $\mathbf{E}$
and $\mathbf{B}$, and making use of the triple vector product operator identity, one arrives at
the differential wave equations for $\mathbf{E}$ and $\mathbf{B}$:

$$\nabla^2\mathbf{E} = \mu\varepsilon\frac{\partial^2 \mathbf{E}}{\partial t^2} \qquad \nabla^2\mathbf{B} = \mu\varepsilon\frac{\partial^2 \mathbf{B}}{\partial t^2}. \tag{2.2}$$

These concise vector expressions describe the propagation of an electromagnetic
wave in a homogeneous, isotropic, linear medium. It can be shown that the constant

multiplying the right-hand side of the wave equation is equal to the square of the velocity of propagation. Thus the velocity of a propagating wave in such a medium is equal to

$$v = 1/\sqrt{\mu\varepsilon}. \tag{2.3}$$

In turn, the index of refraction of the medium is the ratio of the speed of light in a vacuum c to that in the material v:

$$n = c/v = c\sqrt{\mu\varepsilon}. \tag{2.4}$$

Because of our assumptions about the material properties of the medium in which the wave propagates, the ∇^2 operates on each orthogonal component of $\mathbf{E}$ and $\mathbf{B}$, so the vector equations can be broken down into six scalar equations and treated separately. The first two are shown here:

$$\frac{\partial^2 E_x}{\partial x^2} + \frac{\partial^2 E_x}{\partial y^2} + \frac{\partial^2 E_x}{\partial z^2} = \frac{n^2}{c^2}\frac{\partial^2 E_x}{\partial t^2} \quad \text{and} \quad \frac{\partial^2 B_x}{\partial x^2} + \frac{\partial^2 B_x}{\partial x^2} + \frac{\partial^2 B_x}{\partial x^2} = \frac{n^2}{c^2}\frac{\partial^2 B_x}{\partial t^2}. \tag{2.5}$$

The form of the equation is identical for the other components (E_y, B_y, E_z, B_z). Each scalar component of the electric and magnetic fields obeys the scalar wave equation. The solution to the wave equation was known long before Maxwell's time. Each scalar component can be mathematically separated into its spatial and temporal factors. For example,

$$E_x = P_x \times \exp(-ikct/n), \tag{2.6}$$

where P is the spatial component of the electrical field that makes up the optical wave and $k = 2\pi/\lambda$, the propagation constant. Because this field describes behavior for a particular wavelength and no spatial variation is present in the exponent, the time-varying part of the field can be dropped for most analyses. When modeling an optical system, it is the spatial behavior of the electric field that is of primary concern.

The spatial distribution of the electric field satisfies the Helmholtz equation:

$$\nabla^2 P + k^2 P = 0. \tag{2.7}$$

This elegant equation must be cast in a form that permits us to solve problems. To this end, a geometry must be established that corresponds to the practical problems we will face (Fig. 2.1). However, it is not useful to make it too specific. To do so, we would lose the ability to apply it to a large number of situations.

In solving the Helmholtz equation, P is expressed as an integral over the aperture, known as the Kirchhoff integral. The derivation of this solution is somewhat involved and would divert this discussion. Instead we will describe the solution for a geometry that is appropriate for diffractive optics. A simple construction will be

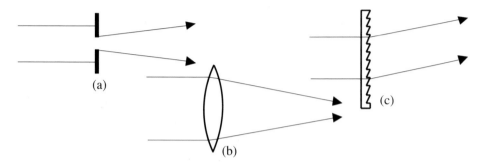

Figure 2.1 A variety of optical phenomena that can be accurately analyzed using scalar diffraction theory. (a) Diffraction from an aperture. (b) Focusing by a thin lens. (c) Diffraction by a grating.

used to introduce the solution. In Fig. 2.2, light traveling in the positive z direction is diffracted by an aperture located in the x–y plane at the origin of the coordinate system. The distribution of the field in that plane is given by $P(x, y, 0)$ and the diffracted field $P(X, Y, Z)$ is evaluated in a plane parallel to the aperture at a distance Z. All points inside the aperture contribute to each point of the diffracted field. To express the field P, we will use the Huygens construction introduced in Chapter 1 to produce the correct relation and give an indication as to the physical sense of the equation.

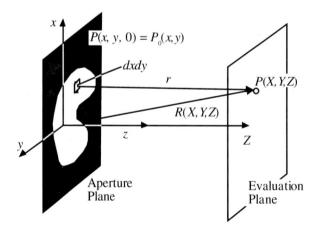

Figure 2.2 The propagation of light from a point $P(x, y, 0)$ within an aperture plane to a point $P(X, Y, Z)$ on an evaluation plane can be calculated using the Rayleigh–Sommerfeld propagation model.

At this point, we can list several more assumptions to simplify the mathematical diffraction models:

(5)   The temporal component of the electric field can be separated and ignored.
(6)   Light is propagating from a source, through a first plane, the aperture plane, where it is defined, to a second plane, the evaluation plane, where its spatial distribution is to be calculated.

(7) The beam size or aperture at the aperture plane is large compared with the wavelength.

(8) The distance from the aperture plane to the evaluation plane is large compared with the wavelength.

Using these assumptions still allows accurate analysis of light propagation from a wide variety of diffractive optical elements. However, it should be noted that assumptions 2 and 7 are typically not valid within a diffractive microstructure. This difficulty is circumvented by determining the complex transmission function of a diffractive element analytically, then applying scalar theory to model the propagation of light to a subsequent plane in the optical system.

With these additional assumptions, Eq. (2.7) can be reduced to the Rayleigh–Sommerfeld integral:

$$P(X, Y, Z) = \frac{1}{2\pi} \iint_{x,y} P(x, y, 0) \frac{e^{-ikr}}{r} \frac{Z}{r} \left(ik + \frac{1}{r} \right) dx dy, \qquad (2.8)$$

where $r = [(x - X)^2 + (y - Y)^2 + Z^2]^{1/2}$ since r is the distance from a point $P(x, y, 0)$ in the aperture plane to a point $P(X, Y, Z)$ on the evaluation plane. Because a diffractive optical element will be located at the aperture plane, we will use a special notation for the field at that point: $P(x, y, 0) = P_0(x, y)$.

This integral [Eq. (2.8)] is a useful tool for analyzing the scalar propagation of light. The eight assumptions listed earlier are valid for a wide range of optics problems. Assumptions 1 to 4 are true for vacuum, air, and most glasses and transparent media; while assumptions 5 to 8 are valid for most cases where a beam of light is propagating from an aperture, lens, or diffractive element to some target plane.

Also, it is a straightforward task to perform these types of calculations on a computer. In the form expressed in Eq. (2.8), the propagation equations are computationally intensive, but a more efficient form will be derived in Sec. 2.2.2 and fully explained in Sec. 3.2. Most of the analyses required for the design of diffractive optical elements can be performed using the Rayleigh–Sommerfeld propagation equation or some simplification thereof. For the set of problems that violates the assumptions of scalar theory, analysis must be based on vector theory, which will be described in Chapter 3.

The Rayleigh–Sommerfeld solution to the wave equation can be further simplified if we make one additional assumption:

(9) The distance from the aperture plane to the evaluation plane is very large compared with the aperture size.

This assumption is sometimes known as the "far-field" approximation. With this assumption in mind, we can return to the derivation of the field in the output plane. We will show that it simplifies to a Fourier transform of the input plane.

Consider that each infinitesimal area of the diffractive optical element in the aperture plane $ds = dxdy$ contributes to the diffracted field P_2 and assume that each of these areas has a field amplitude density $E_u(x, y)$ that produces its own secondary spherical wavelet, e^{ikr}/r, so that the field due to the area ds is

$$dP = \frac{E_u(x, y, 0)}{r} \exp[i(\omega t - kr)]ds, \qquad (2.9)$$

where $r = [(x - X)^2 + (y - Y)^2 + Z^2]^{1/2}$. If the distance from the origin to the evaluation point P_2 is $r = [X^2 + Y^2 + Z^2]^{1/2}$, then the distance between a point in the aperture at $(x, y, 0)$ and a point in the diffracted field at (X, Y, Z) can be written as

$$r = R[1 + (x^2 + y^2)/R^2 - 2(xX + yY)/R^2]^{1/2}. \qquad (2.10)$$

In the far field, because the aperture is located near the origin of the coordinate system and $R \gg x^2, y^2$, the second term is small compared with unity and the other term, so the distance r can be approximated as

$$r \cong R[1 - 2(xX + yY)/R^2]^{1/2}. \qquad (2.11)$$

Because the R^2 term is much larger than xX and yY, only the first two terms in a Taylor's series expansion $[r = R\sqrt{1 - \varepsilon} \approx R(1 - \varepsilon/2)]$ must be retained and Eq. (2.12) becomes

$$r = R - \frac{xX}{R} - \frac{yY}{R} = R - \alpha x - \beta y, \qquad (2.12)$$

where $\alpha = X/R$ and $\beta = Y/R$ are the direction cosines of the point (X, Y, Z) relative to the origin in the aperture plane (Fig. 2.2). Thus, the pattern in the far field will be the same for any evaluation plane; it simply grows linearly in size as the evaluation plane distance Z is moved farther from the aperture.

Based on these approximations, an expression can be found for the diffraction pattern in the evaluation plane based on the field $P_0(x, y)$ in the aperture. Integrating Eq. (2.9) over all of the contributions inside the aperture, we arrive at an expression for the field at the evaluation plane:

$$P(X, Y, Z) = \int_{\text{aperture}} \frac{E_u(x, y, 0)}{r} e^{ikr} dxdy. \qquad (2.13)$$

By inserting the approximation for r [Eq. (2.12)] into this equation and invoking the slow variation of r in the denominator and $r \approx R$, compared with the rapid variations with x and y in the exponent, r can be replaced by R in the denominator to obtain:

$$P(X,Y,Z) = \int_{\text{aperture}} \frac{E_u(x,y,0)}{R} e^{-ikR} \exp[-ik(xX/R)] \exp[-ik(yY/R)] dxdy.$$

(2.14)

There are two further simplifications that will make connections between this integral expression and the Fourier transform. We define two components of the **k** vector:

$$k_X = \frac{kX}{R} = \mathbf{k}\alpha; \quad k_Y = \frac{kY}{R} = \mathbf{k}\beta,$$

(2.15)

where α and β are the direction cosines of the propagation vector **k**. The third component, k_z, has a direction cosine $\gamma = z/R$, but our integration is over only the aperture. The complex amplitude in the aperture plane $u(x,y,0)$ is related to the field amplitude density E_u by

$$u(x,y,0) = \frac{E_u(x,y,0)}{r} e^{-ikr},$$

(2.16)

permitting Eq. (2.14) to be written in a compact and useful form:

$$P(X,Y,Z) = \int_{\text{aperture}} u(x,y,0) \exp[-ik(xX/R)] \exp[-ik(yY/R)] dxdy.$$

(2.17)

It is usually the case for transmitting diffractive optical elements that the aperture includes phase variations, so phase should be included in the function $u(x,y,0)$ as $u_0(x,y) \exp[i\phi_0(x,y)]$. (As was true earlier, the z coordinate has been dropped from further expressions because the diffractive element is located at the aperture plane.) This gives (finally!) a result that is applicable to many diffractive optical elements:

$$P(X,Y,Z) =$$
$$\int_{\text{aperture}} u_0(x,y) \exp\{i\phi_0(x,y) \exp[-ik(xX/R)] \exp[-ik(yY/R)]\} dxdy.$$

(2.18)

This integral is in the form of a two-dimensional Fourier transform and is usually written in a compact notation as $P(X,Y,Z) = \Im\{u_0(x,y) \exp[i\phi_0(x,y)]\}$, which shows that the far-field diffraction pattern $P(X,Y,Z)$ is the Fourier transform of the aperture function, $u_0(x,y) \exp[i\phi_0(x,y)]$.

Another way to express Eq. (2.18) is to write it in terms of the direction cosines α and β. From Eq. (2.15) we see that the evaluation plane coordinates can be written as a function of α and $\beta : X = \alpha R$ and $Y = \beta R$. Because R is the distance between the origin and the point on the evaluation plane, it is essentially constant, so instead of expressing the result in terms of points in the evaluation plane (X, Y), the diffracted field is given in terms of the direction cosines as

$$P(\alpha, \beta) = \int_{\text{aperture}} u_0(x, y) \exp[i\phi_0(x, y)] \exp(-ik\alpha x) \exp(-ik\beta y) dx dy. \quad (2.19)$$

This expression will be useful in Chapter 3, where some of our assumptions, particularly 7 and 8, no longer hold.

What is seen and measured in the laboratory or in an application is not the field P, but the irradiance, which is proportional to the square of the field. But all the heavy lifting in computing the transmission of a diffractive optical element has been done by determining the far-field pattern of its aperture function, which is, as we just noted, the Fourier transform of the aperture function.

2.2 Fourier Analysis

Modeling and understanding optical propagation can be greatly simplified by using the properties of the Fourier transform. As we showed earlier, the far-field diffraction pattern for any aperture and phase can be swiftly calculated using a Fourier transform. Because the transforms of many simple functions are known, a lot can be explained using these known relationships. To illustrate this, we must first introduce three additional concepts: the convolution, the Dirac delta function, and the convolution theorem of Fourier transforms.

Some of the aperture functions consist of a combination of two functions, in which one function is dependent on the coordinates of the other. This combination is called a *convolution*.[2,4] Mathematically, the convolution of two functions $f(x)$ and $g(x)$ is defined as

$$h(X) = \int_{-\infty}^{\infty} f(x) g(X - x) dx, \quad (2.20)$$

which can also be written as $h(X) = f(x) \otimes g(x)$. In effect, the function $f(x)$ multiplies $g(x)$ at all values of x for each value of X and gives a function $h(X)$.

The simplest example is a function convolved with itself [$f(x) = g(x)$]. For example, let $f(x) = \text{rect}(x/b)$ where $\text{rect}(x/b) \cdot 1$ in the range $-b/2 < x < b/2$ and zero elsewhere [Fig. 2.3(a)]. One way of thinking about the convolution of two rect functions is that the convolution consists of the overlap of the areas between the two functions after one of the functions is reversed ($x \rightarrow -x$). Obviously since the functions have widths of one unit, the convolution $h(x)$ will be zero until the centers of the two functions are a distance b apart and will increase until the centers overlap. At that point the area overlap will be unity. In between, the overlap varies linearly with separation [Fig. 2.3(b)].

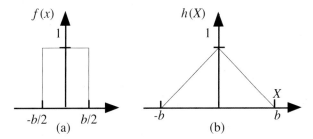

Figure 2.3 Convolution of a rect function with itself. (a) The rect function. (b) The convolution, $h(X) = \text{rect}(x/b) \otimes \text{rect}(x/b)$.

2.2.1 The Dirac delta function

A particularly peculiar (but useful) function for Fourier analysis is the Dirac delta function $\delta(x)$. This function is infinitely narrow and infinitely high at the point where its argument x is zero, but its area is equal to unity.

$$\int_{-\infty}^{\infty} \delta(x) = 1 \quad \text{and} \quad \delta(x) = 0, \quad \text{when } x \neq 0. \tag{2.21}$$

By itself, the Dirac delta function is of little use, but when it is used in a convolution, it permits the generation of functions that would otherwise be very difficult without it. (Also, as we will see, it provides some insight and an approach to the design of diffractive optical elements.) One application of the Dirac delta function that demonstrates its utility consists of expressing two side-by-side apertures in a compact form. They are arranged in what is known in an elementary physics text as Young's double-slit geometry. This arrangement can be represented as the sum of two rectangle functions of width b separated by a distance a: $\text{rect}[(x + \Lambda/2)/b] + \text{rect}[(x - \Lambda/2)/b]$ as shown on the right side of Fig. 2.4. However, it can also be written using a convolution of the rectangle function and two delta functions as $\text{rect}(x/b) \otimes [\delta(x + \Lambda/2) + \delta(x - \Lambda/2)]$ (left side of Fig. 2.4). Although it may seem that the expression using the convolution is longer and more elaborate, this representation will prove to be more useful.

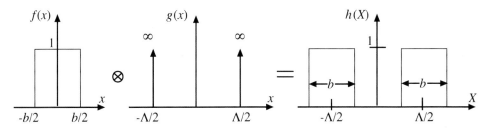

Figure 2.4 Convolution of the rect function with a pair of delta functions to produce a double-slit function.

Another useful function is the comb function, which is a periodic array of delta functions:

$$\text{comb}(x/\Lambda) = \sum_{n=-\infty}^{\infty} \delta(x - n\Lambda). \qquad (2.22)$$

It is called a "comb" function because its plot looks like the regularly spaced teeth of a comb. This function can be used with other functions to easily generate various diffraction gratings. The simplest example of an amplitude grating is the convolution of a rect function with a comb function, $[\text{rect}(x/b) \otimes \text{comb}(x/2b]$, to produce a square-wave grating, as shown in Fig. 2.5.

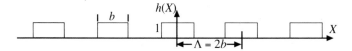

Figure 2.5 Square-wave function formed by the convolution of a rect function and a comb function, where the period of the grating is twice the width of the rect function ($\Lambda = 2b$).

2.2.2 The convolution theorem

These manipulations are of more than passing interest because of the convolution theorem of Fourier theory. The convolution theorem states that the Fourier transform of the convolution of two functions is equal to the product of the Fourier transform of the individual functions. That is,

$$\Im[f \otimes g] = \Im[f] \times \Im[g]. \qquad (2.23)$$

This is an extremely useful result because knowledge or analysis of the individual functions can be used to understand and design optics. Consider the example we just used to illustrate convolution: a single aperture and a pair of delta functions. Because the far-field diffraction pattern is the square of the Fourier transform of this aperture function, the convolution theorem can be used to show very simply that the far-field pattern of a double slit is made up of the sinusoidal fringes of double-slit interference multiplied by that of the single-slit diffraction pattern.

When programming a computer to calculate the effects of an optical component on a propagating wave, it is common to rely on the mathematical properties of the fast Fourier transform (FFT) because of the ease of computation. The Rayleigh–Sommerfeld propagation equation [Eq. (2.8)] has the form of a two-dimensional convolution.[2] We can express the field in the evaluation plane as the convolution of the field at the aperture with a function H. The convolution can be written mathematically as

$$u(x, y, Z) = u_0(x, y) \otimes H(x, y, Z), \qquad (2.24)$$

where

$$H(x, y, Z) = \frac{1}{2\pi} \left[ik + \frac{1}{(x^2 + y^2 + Z^2)^{1/2}} \right) \frac{Ze^{-ik(x^2+y^2+Z^2)^{1/2}}}{(x^2 + y^2 + Z^2)}.$$

and H is the propagation kernel for a defined distance Z. The convolution given as Eq. (2.24) maps the input scalar field $u_0(x, y)$ due to the diffractive optical element, which defines the spatial part of the initial optical wave at the aperture, into the resultant field $u(x, y, Z)$. Thus, $u(x, y, Z)$ is the result of propagating an optical wavefront $u_0(x, y)$ through a distance Z. This convolution approach to solving the Rayleigh–Sommerfeld equation for propagation is called the *plane-wave spectrum* (PWS) method and is fully explained in Chapter 3.

Computationally, wave propagation using a scalar approximation can be very efficiently performed using fast Fourier transform methods with the wavefront treated as a two-dimensional complex array. The entire mathematical operation is performed on this array using standard FFT algorithms.[5] The key issue for simulating scalar propagation using FFTs is sampling of the optical wavefront. The density of data points must adequately sample all the spatial frequencies present in the propagating wave so that an accurate model is generated. For simple cases, e.g., nearly collimated light propagating from an aperture of diameter D, sampling interval Δ and number of points N, the following three rules will ensure accurate modeling.

(1) $\Delta \ll D$. There must be numerous sampling points across the aperture diameter.

(2) $N > (D^2/Z\lambda)$, where Z is the propagation distance. The N points must also adequately sample the output plane.

(3) $\Delta < (Z\lambda/D)$. The sampling interval must adequately resolve spatial frequencies in the output plane.

These rules ensure adequate sampling when performing Rayleigh–Sommerfeld propagations using a computer. Propagation of wavefronts from diffractive structures can be modeled very well using this approach. FFT analysis is used throughout this text for finite wavefront propagation from diffractive structures to output planes.

2.3 Using Fourier Analysis

Diffraction efficiency calculations can be performed for many periodic structures using Fourier transforms.[6] The most straightforward example may be that of a linear blazed grating with depth d and period Λ (Fig. 2.6). (As described in Sec. 1.5, a grating is referred to as "blazed" if it directs a large fraction of the incident light into one of the diffracted grating orders, rather than distributing it over many orders.) This phase grating transmits light with no attenuation, but imparts a phase

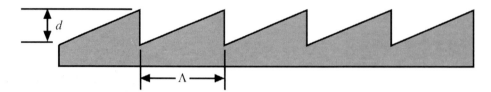

Figure 2.6 Linear blazed grating of period Λ and depth d.

variation across the wavefront. If a planar wavefront is incident on this grating, then the transmitted wavefront (aperture function) can be expressed as a convolution of a comb function with a rect function multiplied by a phase term that is linear in x with a slope ϕ/Λ (we take the amplitude of the field to be unity):

$$U(x) = \text{comb}\left(\frac{x}{\Lambda}\right) \otimes \text{rect}\left(\frac{x}{\Lambda}\right) \exp(i\phi x/\Lambda), \tag{2.25}$$

where the phase difference that is due to the grating depth is given by $\phi = 2\pi(n-1)d/\lambda$ and n is the refractive index of the grating material.

Because the grating is a periodic structure, light propagating away from the surface will resolve itself into discrete diffraction orders whose angles of propagation θ_m are given by the grating equation:

$$\Lambda \sin(\theta_m) = m\lambda. \tag{2.26}$$

In the far field, these diffraction orders are (mathematically) the Fourier transform of U. The angular separation of the diffraction orders is controlled by the grating period, Λ. The light will be diffracted into several diffraction orders, but the diffraction efficiency in each of those orders is determined by the grating depth, d. It can be shown that the fraction of light going into the mth diffraction order is the magnitude squared of the Fourier coefficient p_m for that order:

$$\eta_m = p_m^2 = \left\{\frac{\sin[\pi(k-m)]}{\pi(k-m)}\right\}^2. \tag{2.27}$$

Studying this equation, one sees that all of the light ($\eta = 1$) will be diffracted into a specific diffraction order m if the denominator is zero. This will occur if the grating is etched to an appropriate depth so that $(n-1)d = \lambda$. Then the grating is blazed for the first order, in which $m = 1$. Thus, an evaluation of Eq. (2.27), will result in all of the incident light being diffracted into the first order (i.e., $\eta_m = 1.0$). This holds true as long as the grating depth is small relative to the period. In cases where the depth approaches the period, other effects, such as shadowing (discussed in Sec. 2.5), begin to reduce the efficiency.

2.4 Diffraction Efficiency of Binary Optics

The analysis of the linear blazed grating given here was based on a number of assumptions that simplified the mathematics to the point where we could calculate the grating performance. To achieve this performance, however, one additional condition must be added for actual gratings: you have to be able to make one. For many years, diffractive structures were limited to gratings fabricated using ruling engines, but in the early 1980s, researchers demonstrated the ability to make precise grating structures using photolithography and ion etching.[7–9] These techniques, borrowed from the expanding microelectronics industry, were described briefly in Sec. 1.7. These methods permit the fabrication of complicated surface profiles using several binary masks in conjunction with optical lithographic and etching techniques as discussed in detail in Chapter 7. The efficiency of these multilevel structures depends on the number of masks and the size of the smallest feature that can be produced. Returning to the earlier example of the blazed grating, scalar theory can be used to determine the efficiency of the various approximations to a linear blaze (Fig. 2.6).

2.4.1 The square-wave grating

We begin with the simplest of gratings: a square-wave phase grating (Fig. 2.7). It consists of rectangular sections that look very much the same as those we used for our convolution example (Fig. 2.5), whose length is half the grating period and whose height introduces a phase difference of π between neighboring sections of the transmitted wavefront. The mathematical description of the unit cell is two offset rect functions with a phase constant applied to the second one. The offset is accomplished with the delta function.

$$\text{unit cell} \equiv f(x) = \left[\text{rect}(x/b) \otimes \delta(x)\right] + \left[\text{rect}(x/b) \otimes \delta(x-b)\right]e^{i\phi}, \qquad (2.28)$$

where, as before, the phase difference, $\phi = 2\pi(n-1)d/\lambda$ is set by d, the height of the rectangular section. When d is equal to $\lambda/(2(n-1))$, the phase difference between the two sections of the grating repeat pattern is π radians. Having defined the unit cell, an infinite grating can be specified with one additional convolution:

$$\text{grating} = f(x) \otimes \text{comb}(x/\Lambda). \qquad (2.29)$$

The discrete orders propagate at angles θ_m as defined in Eq. (2.26). Therefore the Fourier transform of the grating is only defined at discrete intervals correspond-

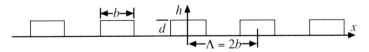

Figure 2.7 Square-wave grating with a period equal to twice the width of the rectangular steps.

ing to the transform of comb(x/Λ), which is comb(Λq), a function with unit value at $q = m/(\Lambda) = m/(2b)$, where m is an integer.

The Fourier transform of the unit cell, $\Im[f(x)] = F(q)$, forms an envelope function that defines the amplitude and relative phase of each diffracted order. The Fourier transform of the rect(x) function is $\sin(q)/q$ or sinc(q), so the transform of the unit cell [Eq. (2.28)] is the sum of two sinc functions with offset phase terms:

$$F(q) = \frac{1}{2}\frac{\sin(\pi bq)}{\pi bq}\exp(i2\pi qb/2) + \frac{1}{2}\exp(i2\pi\phi)\frac{\sin(\pi bq)}{\pi bq}\exp(-i2\pi qb/2).$$

(2.30)

In the case of the ideal binary phase grating, where $\phi = \pi$, this simplifies to

$$F(q) = \frac{1}{2} \times \frac{\sin(\pi bq)}{\pi bq}[\exp(i\pi bq) - \exp(-i\pi bq)].$$

(2.31)

For the case of a square-wave grating, $b = \Lambda/2$ and evaluating for those where the comb function is nonzero, the output of the grating is defined:

$$F(q)|_{q=\frac{m}{\Lambda}} = \frac{\sin(m\pi/2)}{m\pi/2} \times [i\sin(m\pi/2)],$$

(2.32)

which equals zero for all even values of m. For odd values of m, the diffraction efficiency from the grating is given by Eq. (2.32). This equation gives the efficiency values shown in Fig. 2.8.

$$\eta_m = p_m^2 \equiv \left|F\left(\frac{m}{\Lambda}\right)\right|^2 = \left(\frac{2}{m\pi}\right)^2.$$

(2.33)

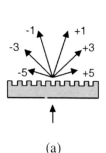

Orders	Efficiency
0	0
+1, −1	40.5%
+2, −2	0
+3, −3	4.5%
+4, −4	0
+5, −5	1.6%
+6, −6	0
(all higher)	3.4%

(a)   (b)

Figure 2.8 (a) A simple binary phase grating (square-wave grating). (b) Table of the efficiencies for each of the orders. The exact 50% duty cycle guarantees the suppression of all the nonzero even orders, and the phase depth of π results in no light diffracted into the zero order.

The table in Fig. 2.8 gives the efficiency for each of the first thirteen orders of a square-wave phase grating.

Note that there is no zero order for this grating. (The argument is fairly simple: think of the grating as pairs of sections, the base section plus the π step. The amount of light transmitted by the step and the base is exactly the same, but they differ in phase by π in the direction of initial light propagation and therefore cancel each other out. Because this happens for each pair of sections, there will be no light propagated in the zero order.) It can be shown that this geometry also results in the suppression of all even orders. The strongest propagation is in the first order, where 40.5% of the light is diffracted into the first order ... and because of symmetry, 40.5% also is directed into the negative first order.

2.4.2 Approximating the blazed grating

To a first approximation, this two-level, or binary, phase grating can be considered as a crude approximation to a blazed grating directing 40.5% of the light into the first order (Fig. 2.9). This binary profile is an equally good approximation of an ideal blaze in the opposite direction because it is also 40.5% efficient in the -1 diffraction order. Thus a linear binary grating with a fixed period can be considered either as an 81% efficient beam splitter or as a 1×2 grating. By modifying this basic profile in a controlled manner, the remaining light can be either diffracted into higher orders or added to the lower orders, depending on the application.

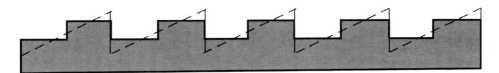

Figure 2.9 Square-wave grating as a rough approximation to a blazed grating.

For example, by using the binary mask approach schematically outlined in Sec. 1.7, the next step would be to etch additional steps that produce $\pi/2$ phase differences. This not only maintains the suppression of the zero order, but also eliminates the negative first-order diffraction and increases the percentage of light directed into the first order to 81%. Additional etching steps using finer and finer masks produce profiles that approach the ideal blazed grating, as shown in Fig. 2.10.

A relatively small number of mask and etch steps can create a binary optic with reasonable efficiency. This leads to the simple equation for first-order diffraction efficiency from an N-level binary optic structure:

$$\eta_1 = \left[\frac{\sin (\pi/N)}{\pi/N} \right]^2. \tag{2.34}$$

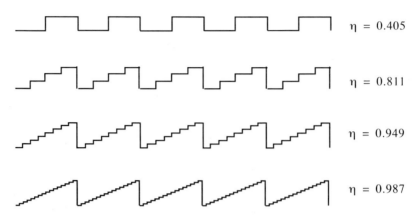

Figure 2.10 Approximations to the profile of an ideal blazed grating by four multilevel gratings. The first-order diffraction efficiency is given to the left of the profiles.

This equation is applicable for any number of levels, provided the steps in the structure are uniform and of optimum depth for diffraction into the first order.

Based on Eq. (2.27), it would appear that diffractive structures should be fabricated with as many masks as possible. However, as the number of levels increases, fabrication issues become a more significant issue. As the feature sizes to be patterned get progressively smaller, the cost and complexity increase. Additional aspects of grating fabrication will be discussed in Chapters 7 and 8.

Implicit in our analysis is the assumption that the structure is infinitesimally thin, since $P_0(x, y)$ was restricted to the $z = 0$ plane. This is valid if the depth d is very small relative to the period, Λ. But as the depth becomes significant, the first-order scalar efficiency calculation becomes less accurate, as we will describe in Sec. 2.5. This analysis can be extended to other types of diffractive elements. While it has been demonstrated for a simple blazed grating, the efficiency calculation can be applied to diffractive structures of widely varied shapes and orientations. The diffractive *pattern* determines the optical wavefront; the *depth and structure* determine the efficiency of the element creating that wavefront.

2.5 Extended Scalar Theory

For any significant diffraction angle, the basic scalar assumptions break down.[10] The analysis in Sec. 2.4.2 assumed that the structure was infinitesimally thin, and that the blazed structure depth was insignificant relative to the grating period. For diffraction angles of more than a few degrees, this assumption is no longer valid. The wavefront emerging from the blazed structure has small gaps introduced by the facets. These gaps are minor discontinuities in the wavefront and are very difficult to detect, but they do affect the diffraction efficiency of the structure. This effect is sometimes referred to as shadowing. For illustration, consider Fig. 2.11. A beam of light, incident from below, has uniform amplitude and phase. As it passes through the diffractive structure, it is redirected and gaps are created in the beam. The small portion of the beam emerging from each period of the grating is a beamlet of width

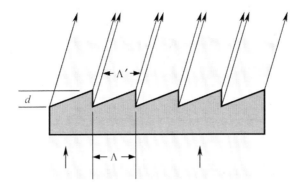

Figure 2.11 Wavefront discontinuities that are due to structure depth. The efficiency is reduced by the amount of the "fill factor," the ratio of the width Λ' of the emerging beamlet to the grating period Λ.

Λ'. Width Λ' is slightly smaller than the period Λ. The "fill factor" on the emerging wavefront is therefore Λ'/Λ.

As before, we can describe the wavefront created by the grating as a complex convolution, this time with a slight perturbation. The comb function is still directly related to the period Λ, but the cell description is now related to Λ':

$$\text{comb}\left(\frac{x}{\Lambda}\right) \otimes \text{rect}\left(\frac{x}{\Lambda'}\right) \exp\left(\frac{i2\pi\kappa'x}{\Lambda}\right). \tag{2.35}$$

The Fourier transform of this function will be a comb function (the diffraction orders) multiplied by a decentered sinc envelope. The diffraction efficiency will be maximized when the sinc envelope is centered on the first diffraction order. This will occur when

$$\frac{\kappa'}{\Lambda} = \frac{\kappa}{\Lambda'}, \quad \text{where } \kappa = (n-1)\frac{d}{\lambda}. \tag{2.36}$$

Because Λ' is smaller than Λ, it follows that there is an optimal grating depth, d', where the sinc envelope is centered on the first diffraction order. This shallower blaze depth will enhance the efficiency of the diffractive structure, sometimes by several percent. From Eq. (2.34) we can express d' as a function of Λ' and Λ, but number of algebra steps will yield a simple equation for optimum blaze depth as a function of the diffraction angle θ_d:

$$d' = \frac{\Lambda'}{\Lambda}d = \frac{\lambda}{n - \cos\theta_d}. \tag{2.37}$$

Even at this optimum grating depth, there is still a significant loss of efficiency. This enhanced scalar efficiency is lower because of the reduced fill factor at the grating. This broadens the sinc envelope in the far field by a factor of Λ/Λ'. The efficiency loss is therefore equal to (Λ'/Λ) (Ref. [2]). When the period is large relative to the structure's depth, this number is approximately 1.0, but as the ratio drops, the efficiency loss predicted by an extended scalar theory becomes significant.

To demonstrate the efficiency losses predicted by extended scalar theory, the diffraction efficiency has been plotted as a function of period/wavelength for a typical glass blazed structure in Fig. 2.12. The plot compares the simple scalar theory prediction (100%), with extended scalar theory and vector theory. The vector theory calculation was done using rigorous coupled-wave analysis (discussed in detail in Chapter 3), solving Maxwell's equations for the air/glass interface. The extended scalar prediction is significantly more accurate than the basic scalar method.

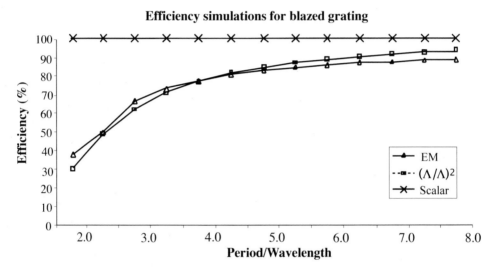

Figure 2.12 Efficiency of a blazed grating as a function of the grating period to wavelength ratio for simulating electromagnetic and extended scalar theory. The standard scalar theory is independent of this ratio and predicts 100% diffraction efficiency for all values.

The efficiency loss for extended scalar theory can be directly applied to stepped diffractive structures. The net efficiency will be that predicted by scalar analysis, multiplied by the extended scalar term. For instance, the scalar efficiency of a four-level grating structure with a period of 4.0 μm, fabricated in a glass of index 1.5 and operating at a wavelength of 1.0 μm would be 0.81 at an optimum depth of 1.88 μm. However, because the diffraction angle is 14.48 deg, the reduced fill factor is 0.878 and the $(\Lambda'/\Lambda)^2$ factor is 0.772. So the net efficiency for this four-level grating structure is 0.772×0.81 or 0.625. Mathematically, this can be represented as

$$\eta_{net} = \eta_{scalar} \times (\Lambda'/\Lambda)^2. \tag{2.38}$$

2.6 Conclusion

In this chapter we have shown that with some reasonable assumptions, a Rayleigh–Sommerfeld analysis can reduce a very general theory of light-wave propagation to a practical method for calculating far-field diffraction patterns. Later on, some of these simplifying assumptions will be removed, enabling one to analyze diffractive structures and effects that violate the far-field approximation. For a large class of diffractive optical elements, Fourier transform theory provides an analytical technique that is useful and easy to understand. Computationally, fast Fourier transforms are the tools that can solve most diffractive optical problems with ease. These Fourier techniques can be used to derive the diffraction efficiencies of diffractive optical elements. In the next chapter we will look at a more rigorous approach to electromagnetic propagation and describe some computations and simplifications that can be used to understand and design certain diffractive elements.

References

1. E. Hecht, *Optics, 4th* ed. Addison-Wesley, Reading, MA (2002).
2. J.W. Goodman, *Introduction to Fourier Optics, 2nd* ed. McGraw-Hill, San Francisco (1996).
3. P. Lorrain and D. Corson, *Electromagnetic Fields and Waves, 3rd* ed. W. H. Freeman, San Francisco (1988).
4. F.T.S. Yu, *Optical Information Processing*. Krieger Publishing, Melbourne, FL (1983).
5. E.A. Sziklas and A.E. Siegman, "Diffraction calculations using fast Fourier transform methods," *IEEE Proc.* **62**, pp. 410–412 (1974).
6. G.J. Swanson, "Binary optics technology: the theory and design of multi-level diffractive optical elements," MIT Technical Report 854, Massachusetts Institute of Technology, Cambridge, MA (1989).
7. G.J. Swanson and W. Veldkamp, "Binary lenses for use at 10.6 micrometers," *Opt. Eng.* **25**(5) (1985).
8. J.R. Leger et al., "Coherent laser beam addition: an application of binary-optic technology," *Lincoln Laboratory J.* **1**(2) (1988).
9. J.A. Cox, "Overview of diffractive optics at Honeywell," *Proc. SPIE* **884** (1988).
10. G.J. Swanson, "Binary optics technology: theoretical limits on the diffraction efficiency of multilevel diffractive optical elements," MIT Technical Report 914, Massachusetts Institute of Technology, Cambridge, MA (1991).

Chapter 3

Electromagnetic Analysis of Diffractive Optical Elements

In Chapter 2, the basis for the electromagnetic behavior of DOEs was introduced. To this end, scalar diffraction theory was presented and formulated from an intuitive point of view. As pointed out in the course of the derivation, there are several approximations inherent in the scalar formulation. For this reason, in certain applications, scalar theory becomes invalid and consequently in these cases a modeling technique that more rigorously accounts for the interaction between the incident optical field and the DOE must be used. In this chapter, we present two such techniques, namely, the modal and the finite-difference time-domain methods (FDTD). Results from these techniques are given, along with a discussion of effective medium theory, a simplified approach to modeling the performance of DOEs with subwavelength features. However, before we introduce these techniques we first present a discussion on the limitations of scalar diffraction theory.

3.1 Scalar Limitations

As presented in Chapter 2, scalar theory treats the interaction between the incident optical field and the DOE as a retardation in the optical path on the incident field, which can be determined from the optical path length through which the field travels. In this case, the interaction is accounted for by adding a phase delay to the incident field that is proportional to the distance of propagation through the DOE profile, as shown in Fig. 3.1. Therefore, as represented in Eq. (2.18), the diffracting field becomes essentially the incident field with an additional position-dependent phase factor. This representation is sometimes referred to as the scalar thin-phase approximation.

Although scalar theory is simple and appealing, it in fact does not rigorously account for the electromagnetic coupling effects along the boundary of the diffracting profile. Unfortunately, for DOEs whose dimensions are comparable to the wavelength of illumination, the scalar approximation introduces errors in the analysis. An example of the effects of small structures can be seen in Fig. 3.2, which shows the diffracting field values evaluated along the boundary of the DOE, over four

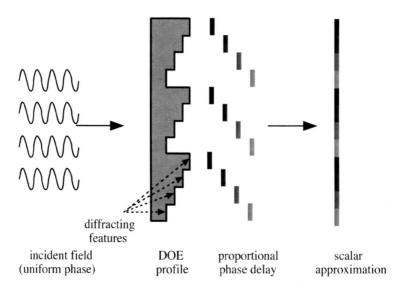

diffracting
features

incident field   DOE   proportional   scalar
(uniform phase)   profile   phase delay   approximation

Figure 3.1 Scalar (thin-phase) approximation for the interaction between an incident field and a diffractive structure.

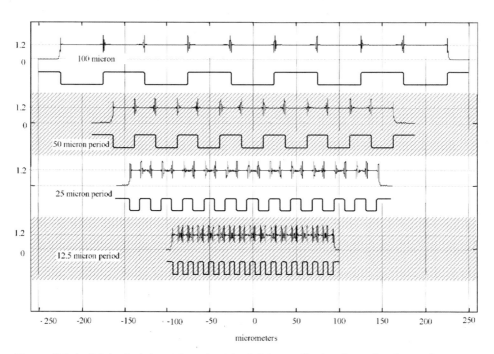

Figure 3.2 Individual plots of the electric field amplitude along the boundary of a DOE, for different periods. The wavelength for each case was 1 µm. The vertical axis represents local values of the electric field amplitude in volts per meter. These results were calculated using the finite-difference time-domain method, which is discussed later in this chapter.

different grating periods, but the same illuminating wavelength, μm. Note that as the relative size of the features in the diffracting profile approaches the wavelength of the optical field, i.e., from top to bottom in Fig. 3.2, the nonuniformities in the boundary field become more pronounced. As a result, for diffracting structures that contain a relatively large number of features on the scale of the wavelength of illumination, errors are introduced by the scalar approximation, rendering it invalid. Clearly, a different modeling approach is needed.

It should be pointed out that the physical basis for the nonuniformities in the electric field along the DOE boundary arise from the increased influence of electromagnetic coupling effects over the boundary, which can only be taken into account by using a more rigorous analysis method. However, if the electric field values are obtained several wavelengths away from the DOE boundary, either experimentally or using a rigorous method, such coupling effects are minimal and the scalar method then becomes a valid method for the propagation of electric field values from one plane to another. Thus, the scalar approach is still useful as part of an overall vector model.

The appeal in using scalar methods in this way stems from issues related to computational efficiency. As will be shown later, the solution of electromagnetic boundary value problems requires significant computational resources. Therefore, by limiting the application of computational methods to the regions where the effects of electromagnetic coupling are significant, such as the DOE boundary, and using more efficient methods to propagate the field values, we compose solution methods that are both highly accurate and computationally efficient. Thus, the scalar methods presented in Chapter 2 can be used in combination with more rigorous methods to analyze DOE performance. Particularly, there exists a scalar model that can be used to propagate field values with improved computational efficiency. It is known as the plane-wave spectrum model.[1]

3.2 Plane-Wave Spectrum Method

The plane-wave spectrum (PWS) method is an accurate method for the propagation of electric field values, provided that the input field values are accurate and the sampling criteria are met. The PWS can be used in conjunction with more rigorous electromagnetic boundary value solvers to efficiently analyze DOE performance.

Because of the linearity of Maxwell's equations, presented in Eq. (2.1), electromagnetic fields can be propagated by superposing the propagated fields from an array of sources located within a plane, as was done when we described Huygens principle in Sec. 1.4. In Chapter 2 scalar diffraction was formulated on this hypothesis and the resulting Rayleigh-Sommerfeld diffraction integral, Eq. (2.15), was derived. Although the use of the RS integral is a viable method for the propagation of electromagnetic fields, it is not the most computationally efficient. This is because the computational dependence of a straight superposition is N^2, which comes from the fact that each source point (which there are N of) is propagated to an array of N locations in the observation plane. Alternatively, the PWS method reduces this dependence to $N \log N$. It does this by exploiting the planar relationship between the output boundary and an observation boundary.

In place of the graphical description of propagating waves using the Huygens construction is a mathematical approach that uses a concept called the *Green's function*. A Green's function is the solution to the wave equation when an ideal point source is used as the driving function. It represents the impulse response for wave propagation in the system being modeled. Thus, according to linear systems theory, one can determine an arbitrary propagating wave by performing a convolution of the various point sources. This can then be used to represent a more general source profile.

If these two boundary planes are parallel to each other, then the Green's function, which governs the behavior of a propagating wave, is shift invariant. As a result, the system of electromagnetic propagation becomes linear and shift invariant. Consequently the propagation from plane to plane can be implemented through a spatial convolution, as described in Sec. 2.2.2. In accordance with standard signal-processing techniques, such a system can be evaluated using the fast Fourier transform, which saves considerable computational time. Therefore in the remainder of this section we present the formulation of the PWS method in terms of its solution using spatial Fourier transforms.

Consider a series of plane waves propagating perpendicularly to the z-axis and at an angle to the x-axis, as shown in Fig. 3.3. The lines represent planes of constant phase separated by one wavelength. For a phase front in the $z = 0$ plane, the propagation vector can be written as $\mathbf{k} = k\alpha\hat{x} + k\beta\hat{y}$, where α and β are the direction cosines defined earlier. We can see that the intersections of the phase fronts along the x-axis have a period T_x. If we denote two consecutive points of intersection as

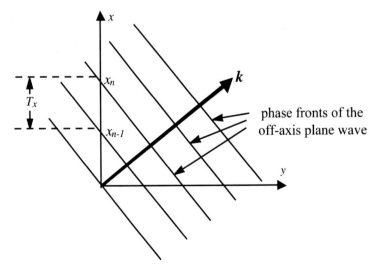

Figure 3.3 Intersection of the phase fronts of a plane wave lying in the x–y plane.

x_n and x_{n-1}, then the distance between them is

$$x_n - x_{n-1} = T_x = \frac{1}{f_x},$$ (3.1)

where the reciprocal of the period T_x is the spatial frequency f_x of the wave along the x-axis.

For a given point x along the x-axis, the phase of the wave is equal to $k\alpha x$. Because the phase difference between consecutive phase fronts is equal to 2π, we can write:

$$2\pi = k\alpha(x_n - x_{n-1}) = \frac{k\alpha}{f_x} = \frac{2\pi\alpha}{\lambda f_x}.$$ (3.2)

Consequently, $k\alpha = 2\pi f_x$ and similarly in the y direction, $k\beta = 2\pi f_y$.

As was shown in Chapter 2, the continuous spatial complex field amplitude $u(x, y)$ can be decomposed into a two-dimensional sum of exponentials see Eq. (2.19):

$$P(\alpha, \beta) = \int_{\text{aperture}} u(x, y) \exp[i\phi_0(x, y)] \exp(-ik\alpha x) \exp(-ik\beta y) dx dy$$ (2.19)

For the purposes of this chapter, it is easier to express this decomposition in terms of the spatial frequencies f_x and f_y, than the evaluation plane coordinates X, Y or their corresponding direction cosines α and β. Substituting for $k\alpha$ and $k\beta$ in the above equation, we can write it in spatial frequency coordinates as

$$P(f_x, f_y) = \int_{-\infty}^{\infty} \int_{-\infty}^{\infty} u(x, y) \exp[-i2\pi(f_x x + f_y y)] dx dy$$ (3.3)

where the exponential terms, $\exp[-i2\pi(f_x x + f_y y)]$, correspond to the Fourier basis in the spatial domain. Note that because these field values exist in two dimensions, the basis set must therefore be two-dimensional (2D). Thus, as is also done in the time domain, we can invert the relation of Eq. (3.3) using an inverse spatial Fourier transform:

$$u(x, y) = \int_{-\infty}^{\infty} \int_{-\infty}^{\infty} P(f_x, f_y) \exp[i2\pi(f_x x + f_y y)] df_x df_y,$$ (3.4)

where $u(x, y)$ is the complex amplitude of the electric field. In this case Eq. (3.4) represents a sum of plane waves, with each wave having a propagation direction of

$$\exp(i\mathbf{k} \cdot \mathbf{r}) = \exp\left[i\frac{2\pi}{\lambda}(\alpha x + \beta y + \gamma z)\right],$$ (3.5)

where **k** is the wave vector that denotes the direction of propagation and $\gamma = (1 - \alpha^2 - \beta^2)^{1/2}$, $\alpha = \lambda f_x$, and $\beta = \lambda f_y$. Recall that the direction of a plane wave can be represented by a wave vector **k** in terms of its position relative to a reference coordinate system, as shown in Fig. 3.4. In this case, the wave vector can be written as

$$\mathbf{k} = \left(k_x^2 + k_y^2 + k_z^2\right)^{1/2}$$

$$= [(k\cos\theta_1)^2 + (k\cos\theta_2)^2 + (k\cos\theta_3)^2]^{1/2}. \tag{3.6}$$

From this notation we can redefine the angles of propagation as $\cos\theta_1 = \alpha$, $\cos\theta_2 = \beta$, and $\cos\theta_3 = \gamma$, where α, β, and γ are referred to as the direction cosines, where $\alpha^2 + \beta^2 + \gamma^2 = 1$, as required.

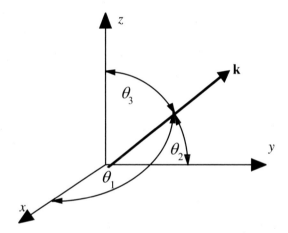

Figure 3.4 Illustration of the propagation of a wave vector **k** relative to the x, y- and z-axes.

Therefore we can represent the plane waves in a real wavefront $u(x, y)$ as an angular spectrum of the direction cosines as

$$A\left(\frac{\alpha}{\lambda}, \frac{\beta}{\lambda}\right) = \int\limits_{-\infty}^{\infty}\int\limits_{-\infty}^{\infty} u(x, y)\exp\left[-i2\pi\left(\frac{\alpha}{\lambda}x + \frac{\beta}{\lambda}y\right)\right]dxdy, \tag{3.7}$$

where $A[(\alpha/\lambda')(\beta/\lambda)]$ is the angular spectrum of the field $u(x, y)$.
    Propagation of $u(x, y)$ to any plane down the z-axis can be achieved by propagating the angular spectrum,

$$A\left(\frac{\alpha}{\lambda}, \frac{\beta}{\lambda}, z\right) = \int\limits_{-\infty}^{\infty}\int\limits_{-\infty}^{\infty} u(x, y, z)\exp\left[-i2\pi\left(\frac{\alpha}{\lambda}x + \frac{\beta}{\lambda}y\right)\right]dxdy, \tag{3.8}$$

or conversely,

$$u(x, y, z) = \int\limits_{-\infty}^{\infty} \int\limits_{-\infty}^{\infty} A(f_x, f_y, z) \exp[i2\pi(f_x x + f_y y)] df_x df_y.$$

Because $u(x, y, z)$ represents the complex amplitude of a propagating wave, it must also satisfy the Helmholtz equation:

$$\nabla^2 u + k^2 u = 0. \tag{3.9}$$

As a result $A[(\alpha/\Lambda'), (\beta/\Lambda'), (z)]$ must satisfy

$$\frac{d^2}{dz^2} A\left(\frac{\alpha}{\lambda}, \frac{\beta}{\lambda}, z\right) + \left(\frac{2\pi}{\lambda}\right)^2 (1 - \alpha^2 - \beta^2) A\left(\frac{\alpha}{\lambda}, \frac{\beta}{\lambda}, z\right) = 0. \tag{3.10}$$

By using the separation of variables technique, i.e.,

$$A\left(\frac{\alpha}{\lambda}, \frac{\beta}{\lambda}, z\right) = A\left(\frac{\alpha}{\lambda}, \frac{\beta}{\lambda}, z\right) A(z), \tag{3.11}$$

the solution to Eq. (3.10) is

$$A\left(\frac{\alpha}{\lambda}, \frac{\beta}{\lambda}, z\right) = A\left(\frac{\alpha}{\lambda}, \frac{\beta}{\lambda}\right) \exp\sqrt{1 - \alpha^2 - \beta^2} z. \tag{3.12}$$

Thus, the field $u(x, y, z)$ in the plane z can be written as

$$u(x, y, z) = \int\limits_{-\infty}^{\infty} \int\limits_{-\infty}^{\infty} A\left(\frac{\alpha}{\lambda}, \frac{\beta}{\lambda}\right) \exp\sqrt{1 - \alpha^2 - \beta^2} z$$

$$\times \exp\left[i2\pi\left(\frac{\alpha}{\lambda}x + \frac{\beta}{\lambda}y\right)\right] d\frac{\alpha}{\lambda} d\frac{\beta}{\lambda}. \tag{3.13}$$

The PWS model represented in Eq. (3.13) provides an efficient formulation for propagating diffracting fields from a source plane to a subsequent observation plane.

Although the PWS is an efficient propagation method, it requires that the electric field values (both magnitude and phase) be specified in the source plane. Consequently, the accuracy of this plane-wave description is solely dependent on the accuracy of the electric field values as determined, or specified, in that plane. Therefore if the field values in the source plane are not valid, the errors introduced will be propagated to the observation plane.

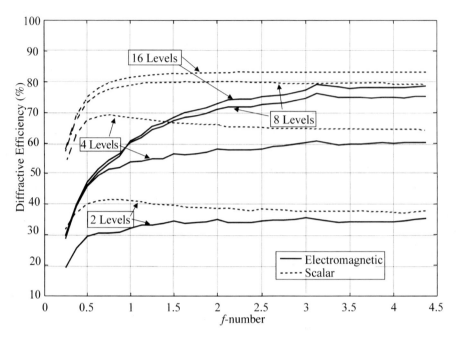

Figure 3.5 Analysis of 3D diffractive lenses as a function of *f*-number using both scalar theory and the finite-difference time-domain method. The vertical axis represents diffraction efficiency, i.e., the amount of energy within the diffraction-limited spot relative to the total energy in the focal plane.

The manifestation of such errors occurs in the assessment of DOE performance. For example, as the *f*-number of a diffractive lens is reduced, the size of the features in the outer zones is also reduced.[2] (The *f*-number is the effective focal length *f* the lens divided by the diameter of its entrance. This is discussed in Sec. 4.1.1.) Thus, for low *f*-number lenses, feature sizes can easily approach the wavelength scale and in some cases become subwavelength. When this occurs, the lens performance departs from the values computed by scalar theory, as shown in Fig. 3.5. To properly account for boundary coupling effects, and thereby determine the actual magnitude and phase profile, a more rigorous computational method must be used. In the remainder of this chapter we discuss two of the more popular methods, namely, rigorous coupled-wave analysis (RCWA), or the modal method (MM), and the finite-difference time-domain methods.

3.3 Electromagnetic Diffraction Models

In principle, Maxwell's equations, the governing equations of electromagnetism, can be used to determine the exact nature of the interaction between any optical field and any diffracting structure. To achieve this goal, however, the mutual interaction within the optical field must be properly accounted for across the boundary of the diffracting structure. This requires that the electromagnetic boundary conditions be enforced at every point along the boundary and that their influence on the behavior of the optical fields be determined. This procedure is otherwise known as solving the electromagnetic boundary value problem. Once the interaction along the boundary is properly accounted for, the resulting field becomes a secondary source. Propagation of this field produces the diffracted fields in the observation plane.

Unfortunately, a closed solution for the electromagnetic boundary value problem can only be obtained for a small number of geometries, namely, cubes, cylinders, and spheres, i.e., those shapes that can be represented or conformally mapped into separable coordinate systems.[3–5] The reason for this is that in order to apply the electromagnetic boundary conditions in a global fashion, one must be able to separate the spatial dependence of the field values along the various coordinate directions. While this can be done for geometries that naturally map onto orthogonal coordinate systems, for most general geometries closed-form analytical solutions do not exist.

Consequently, for the analysis of DOEs, which in general contain highly irregular structures, one must resort to numerical techniques. Currently the two most popular techniques for the electromagnetic analysis of DOEs are the modal method, for infinitely periodic structures, and the finite-difference time-domain method for aperiodic structures. In the remainder of this section we present both methods, beginning with the modal method.

3.3.1 Modal method

The formulation of the modal method (MM) is predicated on the condition that the DOE profile is periodic. The profile can take the form of a periodic volume distribution, periodic surface relief profile, or a periodic surface conductivity. If this condition is met, then the material structure and the electromagnetic fields can be represented in terms of series expansions. In the MM, the field expansions consist of individual modes, where each mode represents a physical solution to Maxwell's equations. If, instead, the fields are represented in terms of a plane-wave expansion, then the solution technique is called the *rigorous coupled* wave analysis (RCWA). An important distinction between the two methods is that the individual plane-wave components used in the RCWA expansion do not represent a solution to Maxwell's equations. Rather, only the entire expansion is a solution. While this distinction results in two different formulations of the problem, they have in fact been shown to be equivalent.[6,7] In the formulation that follows we focus on the modal method.

To begin our formulation of the modal method, we assume a TE polarized field (i.e., the electric field is perpendicular to the plane of incidence) ($E_y = 0$) is incident on an infinitely periodic grating, as shown in Fig. 3.6. In this case the E_y, H_x, and H_z electromagnetic field components are present. We first write down the electric field in each region, apply boundary conditions across the respective regions, and solve the resulting eigenvalue problem.

In region I we have the incident field:

$$E_y^{\text{inc}} = \hat{a}_y \exp(-i\mathbf{k} \cdot \mathbf{r}),$$ (3.14)

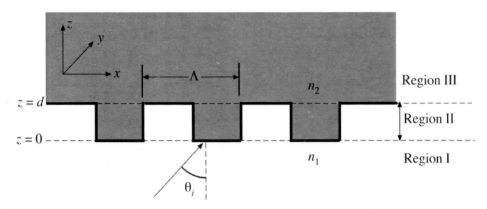

Figure 3.6 Geometry of grating used to formulate rigorous coupled-wave analysis and modal methods.

where the magnitude of the wave number is given by

$$k_0 = \frac{2\pi}{\lambda}. \tag{3.15}$$

To allow for on-axis illumination, the wave number is represented in terms of its projection onto the $\hat{x}$- and $\hat{z}$-axes:

$$\mathbf{k} = \alpha_0 \hat{x} + r_0 \hat{z}, \tag{3.16}$$

where $\alpha_0 = n_1 k_0 \sin \theta_i$ and $r_0 = n_1 k_0 \cos \theta_i$. In a similar fashion, the reflected field is represented as

$$E^r = \sum_m R_m \exp[-i(\alpha_m x - r_m)]$$

$$= \sum_m R_m \exp[-i(\mathbf{k}_m^r \times \mathbf{r})], \tag{3.17}$$

where $\mathbf{k}_m^r = \alpha_m \hat{x} - r_m \hat{z}$. The negative sign of the $\hat{z}$ term represents a component propagating along the negative z direction.

Note that the expansion for the reflected wave is simply a plane-wave expansion, with each component representing a diffraction order of the reflected wave. The reason the basis vectors in this expansion are plane waves is because the modes of a wave propagating in unbounded free space are in fact plane waves. However, this is not the case when waves are subject to inhomogeneities or are confined in one or more dimensions. Such is the case in region II.

Because the diffracting structure is assumed to have periodicity Λ in the x direction, α_m satisfies the Bragg condition:

$$\alpha_m = \alpha_0 + \frac{2\pi m}{\Lambda}. \tag{3.18}$$

Therefore we can represent the z component of the wavenumber as

$$r_m = \begin{cases} [(n_1 k_0)^2 - (\alpha_m)^2]^{1/2} & \alpha_m \leq n_1 k_0 \\ i[(\alpha_m)^2 - (n_1 k_0)^2]^{1/2} & \alpha_m > n_1 k_0. \end{cases} \tag{3.19}$$

Of the two expressions, the bottom one corresponds to nonpropagating (evanescent) waves.

In region III we have a similar expression for the transmitted field:

$$E_y^t = \sum_m T_m \exp{-i[\mathbf{k}_m^t \times (\mathbf{r} - h\hat{z})]}, \tag{3.20}$$

where the $h\hat{z}$ term accounts for the substrate thickness, $\mathbf{k}_m^t = \alpha_m \hat{x} + t_m \hat{z}$, and

$$t_m = \begin{cases} [(n_2 k_0)^2 - (\alpha_m)^2]^{1/2} & \alpha_m \leq n_2 k_0 \\ i[(\alpha_m)^2 - (n_2 k_0)^2]^{1/2} & \alpha_m > n_2 k_0. \end{cases} \tag{3.21}$$

Note that the positive sign for the $\hat{z}$ variable used in $\mathbf{k}_m^t$ indicates forward propagation in the $\hat{z}$ direction, whereas the negative sign for $\hat{z}$ in $\mathbf{k}_m^r$, as we noted earlier, indicates backward propagation in the $\hat{z}$ direction, as is necessary for the transmitted and reflected waves, respectively.

The expression for the fields in region II, the grating region, is complicated, owing to the inhomogeneous nature of the grating structure. Thus the field in that region must be expressed in terms of a modal decomposition,

$$E_y(x, z) = \sum_m E_{ym}(x, z) = \sum_m E_{ym} \exp[i\alpha_m x + \gamma_m z], \tag{3.22}$$

where E_{ym} represents a set of eigenvectors, or modal coefficients, of the individual modes of the electric field inside the grating structure. In addition, γ_m represents a set of eigenvalues to be determined and α_m was defined in Eq. (3.18).

Because each component in Eq. (3.22) represents an allowable mode within the grating, it is a solution to Maxwell's equations. As such, each mode E_{ym}, has a corresponding magnetic field mode, H_{ym},

$$\frac{\partial E_y}{\partial z} = -i\omega\mu_0 H_x + \frac{\partial}{\partial y}\left(\frac{1}{i\omega\varepsilon}\frac{\partial H_x}{\partial y}\right) \tag{3.23}$$

$$\frac{\partial H_x}{\partial z} = -i\omega\varepsilon E_y + \frac{1}{i\omega\mu_0}\frac{\partial^2 E_y}{\partial x^2}. \tag{3.24}$$

Because the electric permittivity in region II is periodic, it can also be represented as a Fourier series:

$$\varepsilon(x) = \varepsilon_0 \sum_p \varepsilon_p \exp\left(\frac{i2\pi p x}{\Lambda}\right) \tag{3.25}$$

$$[\varepsilon(x)]^{-1} = \varepsilon_0^{-1} \sum_p \chi_p \exp\left(\frac{i2\pi px}{\Lambda}\right), \tag{3.26}$$

where ε_p and χ_p represent the Fourier coefficients for the permittivity and its inverse, respectively. Note that although we show the Fourier expansion of the inverse permittivity [Eq. (3.27)], it is not used in the MM formulation for TE polarization. Rather, it is used in the TM formulation (i.e., the magnetic field is perpendicular to the plane of incident). Thus, substituting Eq. (3.22) into Eqs. (3.23) and (3.24) we have:

$$i\gamma_m E_{ym} = -i\omega\mu_0 H_{xm} \tag{3.27}$$

and

$$\frac{\partial H_{xm}}{\partial z} = -i\omega\varepsilon E_y + \frac{1}{i\omega\mu_0}(-\alpha_m^2)E_y. \tag{3.28}$$

Using Eq. (3.27) we can reduce Eq. (3.28) to

$$\gamma_m^2 E_{ym} = k_0^2 \sum_p \varepsilon_{m-p} E_{yp} - \alpha_m^2 E_{ym}, \tag{3.29}$$

which is an eigenvalue equation with γ_m^2 as the eigenvalues and E_{ym} as the eigenvectors. This can be cast as a linear system expressed as a matrix:

$$\gamma \tilde{E}_y = [Z]\tilde{E}_y, \tag{3.30}$$

where $\tilde{E}_y = [E_{y1}, E_{y2}, \ldots, E_{yN}]^T$ and $Z_{mn} = k_0^2 \varepsilon_{m-n} - \delta_{mn}\alpha_m^2$.

Once this eigensystem is solved, we can express the electric field in region II in terms of its modal expansion:

$$E_y(x, y) = \sum_{l=1} \{A_l \exp(i\gamma_l z) + B_l \exp[-i\gamma_l(z - h)]\} \times \sum_m E_{yml} \exp(i\alpha_m x). \tag{3.31}$$

By solving the eigenvalue equation of Eq. (3.29) we have determined the set of modes, i.e., basis vectors that the electric field can sustain in the grating region. Equation (3.31) now represents the actual field profiles that exist in the grating region based on the properties of the incident field. In this way, the unknown coefficients, A_l and B_l, represent the actual magnitude of the electric field in a given mode. As such, they are determined by matching the boundary conditions at the respective interfaces between regions I, II, and III.

The boundary conditions are applied at the bottom, $z = 0$, according to:

$$E_y^I(x, 0) = E_y^{II}(x, 0)$$
$$H_x^I(x, 0) = H_x^{II}(x, 0), \tag{3.32}$$

which produces

$$\sum_l [A_l + B \exp(j\gamma_l h)] E_{yml} = T_m$$

$$k_0 \sum_l [A_l - B \exp(j\gamma_l h)] H'_{xml} = -t_m T_m, \tag{3.33}$$

where $H'_{xml} = H_{xml}/\eta_0$ and η_0 is the impedance of free space, 377 Ω.

Similarly, the boundary conditions at the interface between regions II and III, $z = h$, are

$$E_y^{II}(x, h) = E_y^{III}(x, h)$$

$$H_x^{II}(x, h) = H_x^{III}(x, h), \tag{3.34}$$

which produces

$$\sum_l [A_l \exp(j\gamma_l h) + B_l] E_{yml} = T_m$$

$$k_0 \sum_l [A_l \exp(j\gamma_l h) - B_l] H'_{xml} = -t_m T_m. \tag{3.35}$$

We can now represent Eqs. (3.33) and (3.35) as a linear system with four equations and four unknowns, in which case we can solve for the coefficients: A_l, B_l, R_m, and T_m. The diffractive efficiency of the grating structure is then determined from R_m and T_m as

$$\eta_{R_m} = \text{Re}\left(\frac{r_m}{r_0}\right) \times |R_m|^2$$

$$\eta_{T_m} = \text{Re}\left(\frac{t_m}{r_0}\right) \times |T_m|^2. \tag{3.36}$$

As an example of this method, we have applied it to the analysis of the square-wave grating discussed in Sec. 2.4.1. The scalar theory efficiencies were listed in Fig. 2.8, and are independent of wavelength. In the present case, the grating structure is fabricated in fused silica that has an index of refraction at 1.457 and we assume that it is illuminated with a helium-neon laser at a wavelength of 0.633 μm. Here the grating performance is computed using the modal method for gratings of 1, 2, 5, and 10 μm. The resulting diffraction orders are shown in Fig. 3.7.

As stated in the introduction to this section, the modal method is applicable only to periodic structures. For the analysis of aperiodic structures using electromagnetic methods, one must use an alternative method. One such method, based on the finite-difference time-domain method, is discussed and formulated next.

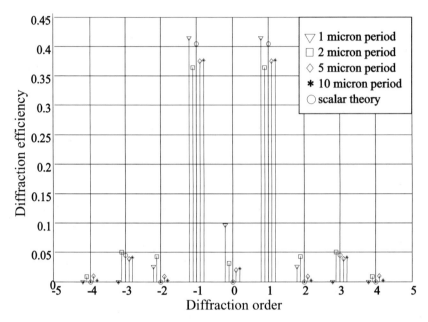

Figure 3.7 Graph showing the grating efficiencies for a square-wave grating fabricated in fused silica with an index of refraction of 1.457 for a range of periods using the modal method. Also shown are the scalar efficiencies for the grating.

3.3.2 Finite-difference time-domain method

From first occurrence the FDTD method imposes a grid on the source region, applies Maxwell's time-dependent curl equations at each point, and iteratively computes their solution for the electric and magnetic field components at all points contained in the computational space. As such, the region of electromagnetic interaction, i.e., the computational region, is subdivided into a set of computational cells, as shown in Fig. 3.8. A typical cell spacing is $\lambda/20$, where λ is the wavelength in the medium. As the FDTD computations are performed, the electrical and/or magnetic fields in any cell are related to the surrounding magnetic field and/or electrical fields in that cell through the difference form of Maxwell's curl equations. The material parameters of the medium are also incorporated in the computation of the field components within each computational cell. To this end, the FDTD method is able to easily deal with the interaction between electromagnetic fields and DOEs having complex shapes and/or inhomogeneous medium. Here we review the important points in the formulation of the FDTD method. However, to better understand the formulation of the FDTD method, we first introduce the notation used in the central difference approximation to a continuous derivative.

The FDTD method derives its name from a direct central finite-difference approximation to these equations. In the formulation of the FDTD method we assume that $f(x, y, z, t)$ is a component of either the electric or magnetic field in an orthogonal coordinate system (x, y, z). We denote discrete points in space as $(i\Delta x, j\Delta y, k\Delta z)$ and discrete points in time as $n\Delta t$. Consequently, $f(x, y, z, t)$ can

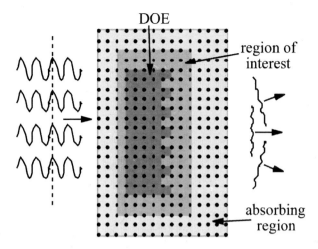

Figure 3.8 Illustration of the computational space for application of the FDTD method to the analysis of diffractive optical elements.

be expressed as

$$f(x, y, z, t) = f(i\Delta x, j\Delta y, k\Delta z, n\Delta t) = f_{i,j,k}^n \tag{3.37}$$

where $\Delta x, \Delta y, \Delta z$ are the lattice increments in x, y, and z coordinate directions, respectively, and Δt is the time increment.

In the FDTD, the central finite-difference expressions for the space and time derivatives, which are accurate to the second order in time and space, are used to approximate the continuous derivatives of Maxwell's equations. For example, consider the first partial space derivatives of f in the u direction calculated at a fixed time $t_n = n\Delta t$:

$$\frac{\partial f(i\Delta x, j\Delta y, k\Delta z, n\Delta t)}{\partial x} = \frac{f_{i+1/2,j,k}^n - f_{i-1/2,j,k}^n}{\Delta x} + O\big[(\Delta x)^2\big]. \tag{3.38}$$

We note the $\pm 1/2$ increment in the i subscript of the x coordinate, as a finite difference over $\pm 1/2\Delta x$. This notation makes the electric and magnetic fields interleaved in the space lattice at intervals Δx. Similarly, the expression for the time-dependent partial derivative of f, evaluated at a fixed point (i, j, k), is

$$\frac{\partial f(i\Delta x, j\Delta y, k\Delta z, n\Delta t)}{\partial t} = \frac{f_{i,j,k}^{n+1/2} - f_{i,j,k}^{n-1/2}}{\Delta t} + O\big[(\Delta t)^2\big], \tag{3.39}$$

where the $\pm 1/2$ increments in the superscript (time coordinate) of f denote a time finite difference over $\pm 1/2\Delta t$. Using these expressions [Eqs. (3.38) and (3.39)], the electric and magnetic fields are then calculated at intervals of $1/2\Delta t$ in accordance with the FDTD algorithm. In the formulations that follow, these equations, called the *central difference equations*, are used to describe the FDTD method for the two-dimensional case.

3.3.2.1 Two-dimensional formulation

In applying the FDTD method to DOE analysis, we first consider Maxwell's time-dependent curl equations for a linear, isotropic, and source-free region,

$$\mu\frac{\partial \mathbf{H}}{\partial t} = -\nabla \times \mathbf{E}$$

$$\varepsilon\frac{\partial \mathbf{E}}{\partial t} = \nabla \times \mathbf{H} + \sigma\mathbf{E},$$

(3.40)

where $\mathbf{E}$ and $\mathbf{H}$ are time-variable electric and magnetic field vectors and μ, ε, and σ characterize their permittivity, permeability, and conductivity, respectively. In two dimensions one can assume either TE or TM polarization. This results from the fact that in two dimensions neither the field nor the DOE profile has any variation in the z direction; consequently all of the partial derivatives with respect to z are zero. Therefore Maxwell's equations can be reduced to two decoupled sets of equations in a rectangular coordinate system,

$$\frac{\partial E_z}{\partial t} = \frac{1}{\varepsilon}\left(\frac{\partial H_y}{\partial x} - \frac{\partial H_x}{\partial y} - \sigma E_z\right)$$

$$\frac{\partial H_x}{\partial t} = -\frac{1}{\mu}\frac{\partial E_z}{\partial x}$$

$$\frac{\partial H_y}{\partial t} = \frac{1}{\mu}\frac{\partial E_z}{\partial x}$$

(3.41)

and

$$\frac{\partial E_x}{\partial t} = \frac{1}{\varepsilon}\frac{\partial H_z}{\partial y}$$

$$\frac{\partial E_y}{\partial t} = -\frac{1}{\varepsilon}\frac{\partial H_z}{\partial x}$$

$$\frac{\partial H_z}{\partial t} = \frac{1}{\mu}\left(\frac{\partial E_y}{\partial x} - \frac{\partial E_x}{\partial y}\right).$$

(3.42)

The first set includes H_x and H_y as components of the magnetic field in the x and y directions, respectively, and E_z as the component of the electric field in the z direction. The second set includes E_x and E_y as components of the electric field in the x and y directions, respectively, and H_z as the component of the magnetic field in the z direction. Applying the central difference expressions [Eqs. (3.39) and (3.40)], the equations for the TE mode are derived to be the following:

$$E_x^{n+1/2}(i,j) = C_a E_x^{n-1/2}(i,j) + C_b[H_z^n(i,j+1) - H_z^n(i,j)]$$

$$E_y^{n+1/2}(i,j) = C_a E_y^{n-1/2}(i,j) + C_b[H_z^n(i+1,j) - H_z^n(i,j)]$$

$$H_z^{n+1}(i,j) = H_z^n(i,j) + [E_x^{n+1/2}(i,j) - E_x^{n+1/2}(i,j-1)$$
$$- E_y^{n+1/2}(i,j) + E_y^{n+1/2}(i-1,j)],$$

(3.43)

where C_a and C_b are material-dependent coefficients and are defined according to

$$R = \frac{\Delta t}{2\varepsilon_0}, \quad R_a = \frac{\Delta t}{\delta^2 \mu_0 \varepsilon_0}$$

$$R_b = \frac{\Delta t}{\mu_0 \delta}, \quad \tilde{E} = R_b E$$

$$C_a(m) = \frac{1 - R\delta/\varepsilon_r(m)}{1 + R\delta/\varepsilon_r(m)}$$

$$C_b(m) = \frac{R_a}{\varepsilon_r(m) + R\delta'}$$

(3.44)

for $m = (i,j)$ and $\delta = \Delta x = \Delta y$.

The electric field component is updated at time $t = n\Delta t$ and the magnetic field components are updated at time $t = (n + 1/2)\Delta t$. The spatial field components are evaluated on an interleaved grid as shown in Fig. 3.9. The difference expressions for the TM mode can be determined in a similar fashion. Once the electromagnetic fields have reached a steady-state condition, they can be propagated to any observation point using more efficient electromagnetic methods, as discussed in the Sec. 3.2.

Both the modal and the FDTD methods represent ways to rigorously solve the electromagnetic boundary value problem for diffractive optical elements. As stated in the first section of this chapter, this becomes necessary when the scale of the features in the DOE profile approaches the wavelength of illumination. These methods can also be used when the feature sizes are subwavelength. However, when the feature sizes are much smaller than a wave, an alternative method can be used. This method is referred to as effective medium theory (EMT).

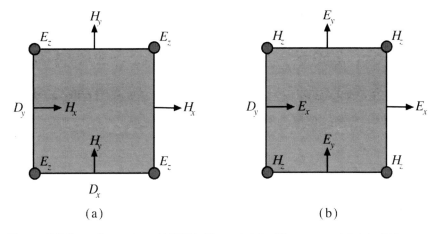

(a)   (b)

Figure 3.9 Two-dimensional FDTD lattices (a) for TE case and (b) for TM case.

3.4 Effective Medium Theory

Effective medium theory describes the interaction between an incident field and a subwavelength structure.[8] It permits one to substitute an "effective" material with homogeneous properties for a spatially varying interface between two materials, as shown in Fig. 3.10. The effective material properties are determined by approximating the electric field within a subwavelength period, Δ, as a constant and applying electromagnetic boundary conditions to the fields across the material interface in order to determine effective material properties.

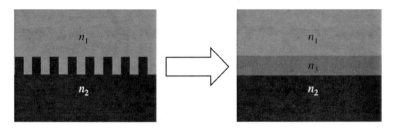

Figure 3.10 Heterogeneous material interface of a subwavelength grating and its effective homogeneous material properties.

The effective material properties for a subwavelength dielectric interface are

$$\varepsilon_{\text{eff}} = \varepsilon_s f + \varepsilon_0 (1 - f) \text{ TE polarization}$$

$$\frac{1}{\varepsilon_{\text{eff}}} = \frac{1}{\varepsilon_s} f + \frac{1}{\varepsilon_0} (1 - f) \text{ TM polarization,}$$

(3.45)

where ε_s and ε_0 are the permittivities of the substrate material and vacuum, respectively; f is the fill factor of the grating profile, of which one period is shown in Fig. 3.11; and $n_{\text{eff}} = \sqrt{\varepsilon_{\text{eff}}}$. This approximation is valid if the subwavelength period, $\Delta \leq \lambda/2n$, where n is the index of refraction of the material and λ/n is the corresponding wavelength in the material. If this condition is not met, then

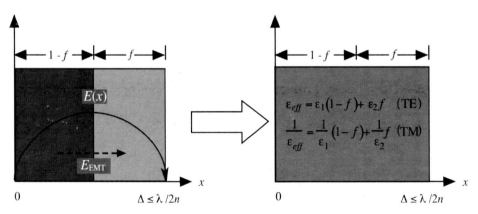

Figure 3.11 Electric field approximation within a subwavelength period as used in effective medium theory.

more than half of a wavelength exists within a subwavelength period. In this case, the constant electric field assumption becomes invalid and more rigorous methods such as those presented earlier must be used.

An alternative approach to the design of subwavelength gratings, developed by Farn,[9] is an eigenfunction expansion method. In this technique, the subwavelength profile is decomposed into a Fourier series expansion that has unknown coefficients. The coefficients represent the start and stop transition points for a subwavelength feature, i.e., its width. The desired continuous profile is also decomposed into a Fourier series with known coefficients. The subwavelength expansion coefficients are determined by forming a system of equations between the two expansions. In general, this system of equations is difficult to solve; however, Farn found an approximate solution for a linear blazed grating,[9]

$$\alpha_i = \frac{i}{N+1} \text{ and } \beta_i = \frac{i}{N}, \tag{3.46}$$

where α_i and β_i are the start and stop transition points for the ith feature and $\beta_N = 1$ is the normalized grating period, as shown in Fig. 3.12. In this approach the number of features, N, is determined by dividing a single period of width W into subperiods of width Δ, $N = W/\Delta$, where $\Delta \leq \lambda/2n$ represents the subwavelength criterion.

In Chapter 11 we present a number of applications of subwavelength diffractive profiles. For more applications on subwavelength technology, see Ref. [10].

In this chapter we have introduced several methods for the rigorous analysis of diffractive optical elements. We presented the formulations and application of the plane-wave spectrum method, the modal method, and the finite-difference time-domain method. While there is a host of other methods one could chose from, by and large these methods represent those that are more commonly used in the literature. In addition, we presented several types of devices that are based on their rigorous behavior, such as subwavelength devices. As fabrication methods continue

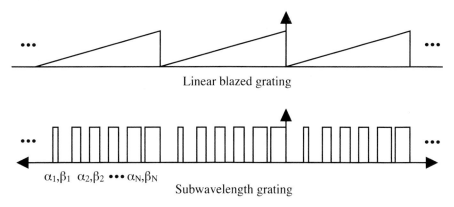

Figure 3.12 Subwavelength design of a linear blazed grating using an eigenfunction expansion of the transition points of the subwavelength profile.

to improve, such elements are becoming increasingly popular in their application, as noted in literature. In the next chapter we build on these methods by discussing the various aspects of diffractive lens design.

References

1. J.W. Goodman, *Introduction to Fourier Optics*, 2nd ed. McGraw-Hill, San Francisco (1996).
2. G.J. Swanson, *Binary Optics Technology: Theoretical Limits on the Diffraction Efficiency of Multi-level Diffractive Optical Elements*, MIT Laboratory Report 854, MIT, Cambridge, MA (1991).
3. H.C. Van de Hulst, *Light Scattering by Small Particles*. Dover, New York (1982).
4. C.A. Balanis, *Advanced Engineering Electromagnetics*. Wiley, New York (1989).
5. A. Ishimaru, *Electromagnetic Wave Propagation, Radiation, and Scattering*. Prentice-Hall, Englewood Cliffs, NJ, pp. 160–161 (1991).
6. T.K. Gaylord and M.G. Moharam, "Analysis and applications of optical diffraction by gratings." *Proc. IEEE* **73**, pp. 894–937 (1985).
7. T.K. Gaylord and M.G. Moharam, "Planar dielectric grating diffraction theories." *Appl. Phys. B* **28**, pp. 1–14 (1982).
8. M. Born and E. Wolf, *Principles of Optics*. Pergamon, New York (1980).
9. M.W. Farn, "Binary gratings with increased efficiency." *Appl. Opt.* **31**, pp. 4453–4458 (1992).
10. J.N. Mait and D.W. Prather, Eds., *Selected Papers on Subwavelength Diffractive Optics*. SPIE Milestone Series, Vol. MS 166. SPIE Press, Bellingham, WA (2001).

Chapter 4

Diffractive Lens Design

This chapter describes the uses of diffractive optics in lens design. In some applications an optical component may require a diffractive surface combined with a classic lens element. In other cases the requirements can be satisfied with just a diffractive element. Both types are described and analyzed here. After describing the concepts and approaches used to design a classic lens, we demonstrate how they can be applied to a single diffractive lens. The use of diffractive surfaces in hybrid lenses to correct chromatic aberration is also described. In Chapter 10, the use of diffractive optics in lens design is extended to multielement systems and to their use in thermal compensation of lenses.

4.1 Basics of Lens Design

The lens design process is depicted in Fig. 4.1. After determining the specifications of the optical system, a starting lens design is evaluated by entering its description in a lens design program. The specifications of the performance of the component as a linear set of weighted target values, called a *merit function*, are set. Then some of the parameters of the optical system are varied and the merit function is evaluated. As the actual parameters approach the desired values, the merit function will approach zero. After a number of iterations, the design has improved (optimized) in comparison with the starting point. It is up to the designer to determine if the performance is good enough for the desired application or if some design changes need to be made, with additional optimization steps.

4.1.1 Describing an optical system

Two sets of data are needed for any lens: the lens data, called the *prescription*, and its surroundings, the *specifications* (Fig. 4.2). The prescription consists of the radius of curvature of each surface plus any additional surface descriptions, the thickness of the medium to the next surface, the refractive index of the medium, and the clear aperture radius of the surface. The specification consists of the object's location, the height of various points on the object (called *field points*), the location of the aperture that limits the size of the light bundles information into the system

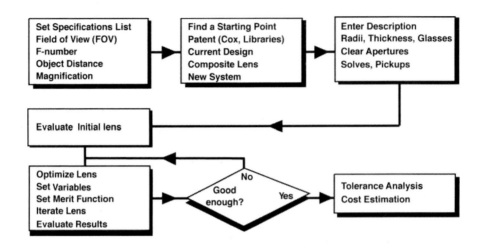

Figure 4.1 The lens design process.

(called the *aperture stop*), the image of the aperture stop as seen from the object (termed the *entrance pupil* of the system), and the wavelengths at which the system is operating. Once the data are entered in a computer, an optical design program traces rays from the specified points through the prescribed lens and determines a *solution*, which consists of the location and size of the image (first-order quantities) plus the departures of the image from perfection (aberrations) at each of the field points.

The specifications for an optical system are derived from its application. In addition to the parameters already described, they specifications usually include some measure of the resolution of the lens (spot size or modulation transfer function),

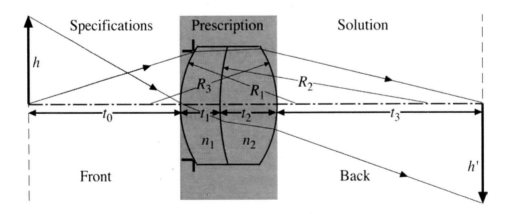

Figure 4.2 A description of an optical system consists of the lens prescription and its surroundings. A ray-trace program determines a solution based on this description.

the size of the object or the angular field of view, and the physical size of the lens. One of the most important specifications of a lens is its f-number, which is defined as the effective focal length of the lens divided by the diameter of its entrance pupil, which is the apparent clear aperture of the lens as seen from the input space. The f-number is a measure of the light-collecting ability of the lens. In the absence of aberrations, it determines the ultimate resolution of the lens. Several lenses with the same focal length and different f-numbers are shown in Fig. 4.3.

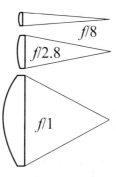

Figure 4.3 Comparison of lenses with different f-numbers.

Current optical design programs not only analyze a design, but also optimize the design when they are provided with measures of the desired performance. The lens is then "corrected" by the program to provide the best performance for the job at hand. This is done by permitting some of the lens parameters to change, while maintaining certain attributes, such as focal length and f-number. Next we describe a simple example of this by correcting chromatic aberration with thin lens descriptions. Later, when diffractive lenses are introduced, this approach will permit us to appreciate their use in combination with conventional (refractive) lenses to correct color aberrations.

4.1.2 The lensmaker's equation

The geometry of a simple refractive lens element is shown in Fig. 4.4. The focal length of a single lens with a center thickness t, refractive index n, and radii of curvature R_1 and R_2 can be expressed as

$$\frac{1}{f} = (n-1)\left[\frac{1}{R_1} - \frac{1}{R_2} + \frac{t(n-1)}{nR_1R_2}\right]. \tag{4.1}$$

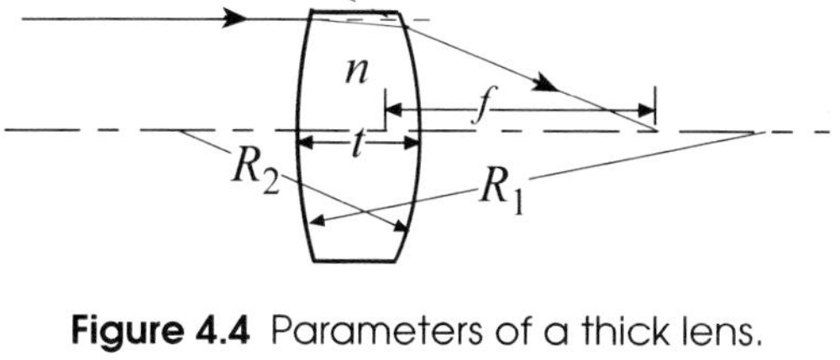

Figure 4.4 Parameters of a thick lens.

If the lens thickness is small compared with its focal length ($t \ll f$), then it can be approximated as a thin lens, one whose thickness need not be taken into account. This permits the third term in Eq. (4.1) to be dropped and the expression reduces to the lensmaker's formula:

$$\frac{1}{f} = (n - 1)\left(\frac{1}{R_1} - \frac{1}{R_2}\right). \tag{4.2}$$

The lensmaker's formula may be compactly written by making two additional definitions. The reciprocal of focal length, $\phi = 1/f$, is the optical power of the lens, and the reciprocal of the radius of curvature, $c = 1/R$, is the curvature of the surface. Making these substitutions gives

$$\phi = (n - 1)(c_1 - c_2) = (n - 1)c, \tag{4.3}$$

where $c = (c_1 - c_2)$ is the net curvature of the lens.

Because some combinations of n, c_1, and c_2 work better than others, a systematic exploration of curvatures in lens design provides a means of optimizing a lens. This is equivalent to changing the shape of the lens and is referred to as lens bending. All of this optimization can be done inexpensively on a piece of paper or in a computer.

In the case of a plano-convex lens ($c_2 = 0$; $R_2 = \infty$), the power and focal length can be written as

$$\phi = (n - 1)c_1; \quad f = R_1/(n - 1). \tag{4.4}$$

These simple equations will be useful for comparing conventional optical elements with their diffractive counterparts and for correcting color errors in an optical system.

4.1.3 Chromatic aberration

Once the prescription for the lens has been entered and the focal length, entrance pupil, and therefore the f-number are determined, the performance of the lens must be evaluated to see if it is suitable for the job at hand. One first-order error is chromatic aberration, which describes the variation of focal length with wavelength. We will now consider the cause of chromatic aberration, followed by a discussion of methods for reducing this effect in a lens system.

Along with the curvature of the surface, the refractive index of a glass is responsible for bending a light ray in a lens. As is evident in Eq. (4.3), the power of a lens is linearly dependent on the refractive index. But the refractive index is a property of the glass and, as can be seen in Fig. 4.5, it is a function of the wavelength of the light passing through it. This variation of refractive index with wavelength is called the *dispersion* of the glass.

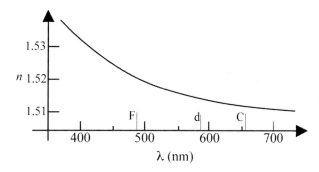

Figure 4.5 Dispersion curve for a glass. The wavelengths labeled F, d, and C on the curve represent special wavelengths that are useful for quantifying the amount of dispersion.

The amount of dispersion of a material can be characterized in number of ways, but the most common expression is that of the Abbe V-number, an expression based on the refractive indices at three wavelengths across the region of interest:

$$V = \frac{n_d - 1}{n_F - n_C}. \tag{4.5}$$

In the visible region of the spectrum, the center, long, and short wavelengths are
$\lambda_d = 589.67$ nm, the yellow line of helium,
$\lambda_C = 656.3$ nm, the red line of hydrogen,
$\lambda_F = 485.1$ nm, the blue line of hydrogen.
The glasses with smaller V-numbers are more dispersive because the denominator in Eq. (4.5) is larger. For the standard assortment of glasses, their V-numbers, based on these wavelengths, range from 70 for the weakly dispersive crowns to 25 for strong flints. Although this concept is applicable in any region of the electromagnetic spectrum, in all of the examples given here these visible wavelengths (C, d, F) will be used. This will permit easy comparisons between the V-numbers of conventional glasses and diffractive structures. However, it should be understood that this same approach could be used across any wavelength band.

We can use the compact form of the lensmaker's formula [Eq. (4.3)] to express the power of a lens at each of these wavelengths:

$$\phi_d = (n_d - 1)(c_1 - c_2) = (n_d - 1)c$$
$$\phi_C = (n_C - 1)(c_1 - c_2) = (n_C - 1)c \tag{4.6}$$
$$\phi_F = (n_F - 1)(c_1 - c_2) = (n_F - 1)c.$$

A measure of chromatic aberration is the difference in powers between the long and short wavelengths: $\Delta\phi = \phi_C - \phi_F = (n_C - n_F)c$. When this chromatic aberration is divided by the power for the center wavelength, the result can be expressed simply:

$$\frac{\Delta\phi}{\phi} = \frac{n_F - n_C}{n_d - 1} = \frac{1}{V_d}. \tag{4.7}$$

Thus the dispersion of a lens material results in chromatic aberration. Because the refractive index is larger at shorter wavelengths, the power of the lens is greater and the focal length is shorter than at longer wavelengths. This means, for example, that blue light will focus at a point closer to the lens than red light, as shown in Fig. 4.6. We will see how this can be corrected by combining lenses of different dispersions or by the addition of a diffractive surface.

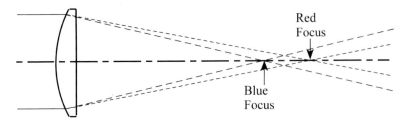

Figure 4.6 Chromatic aberration of a single lens.

4.1.3.1 Correcting chromatic aberration

Chromatic aberration can be corrected by combining lenses whose total power is equal to the required power, but whose total dispersion is zero. For example, given two lenses in contact (Fig. 4.7) whose optical powers, ϕ_a and ϕ_b, sum to the total power of the final lens ϕ:

$$\phi = \phi_a + \phi_b = (n_a - 1)c_1 + (n_b - 1)c_2, \tag{4.8}$$

where n_a and n_b are the refractive indices at the central wavelength λ_d for the two lenses ($n_a = n_{d_a}$; $n_b = n_{d_b}$) and c_1 and c_2 now refer to the sum of the curvatures of each of the individual lenses, $c_a = (c_1 - c_2)_a$ and $c_b = (c_1 - c_2)_b$. The refractive indices and powers of the individual lenses are chosen so that the sum of the individual chromatic aberrations of the lenses cancel:

$$\Delta\phi = \Delta\phi_a - \Delta\phi_b = 0, \tag{4.9}$$

where the cancellation of the individual chromatic aberration is expressed as

$$\Delta\phi = (n_F - n_C)_a c_a + (n_F - n_C)_b c_b = 0. \tag{4.10}$$

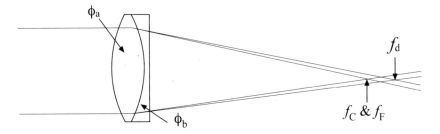

Figure 4.7 Correction of chromatic aberration.

But $c_a = f_a/(n_a - 1)$ and $c_b = f_b/(n_b - 1)$. Inserting these lens curvatures into Eq. (4.10) gives

$$\frac{(n_F - n_C)_a}{n_{d_a} - 1}\phi_a = -\frac{(n_F - n_C)_b}{n_{d_b} - 1}\phi_b, \text{ or } \dots$$

$$\frac{\phi_a}{V_a} = \frac{-\phi_b}{V_b}. \tag{4.11}$$

Since the V-numbers of all glasses are positive, the powers of the lenses must be of opposite sign. So an achromatic doublet is made up of one positive lens and one negative lens. Although the powers (and focal lengths, f_C and f_F) of the doublet at the long and short wavelengths are the same, they are different from the power (and focal length) of the central wavelength, f_d in Fig. 4.7. This type of chromatic aberration is known as secondary color. We will see in Sec. 4.4.1 that diffractive surfaces can be used to provide color correction in place of additional lenses.

4.1.4 Third-order errors

In addition to color errors, there a number of other errors that degrade the image produced by an optical system. These errors tend to spread out ray bundles that originate from an object point. It is possible to classify these errors in terms of the point of origin of the rays in the object plane and where they are directed into the optical system. The lowest order of errors are the Seidel aberrations, or third-order aberrations: spherical aberration S_I, coma S_{II}, astigmatism S_{III}, field curvature S_{IV}, and distortion S_V. These errors are discussed further in Chapter 10. In this section we examine the first three: spherical aberration, coma, and astigmatism.

Spherical aberration is an error that occurs because rays at the edge of a bundle imaged by a lens do not focus to the same point. For example, in Fig. 4.8 the rays that enter at the top and bottom of the lens (called *marginal rays*) focus closer to the lens (at the marginal focus) than rays that propagate close to the axis. The latter rays focus at the paraxial image plane. One measure of spherical aberration is the difference between these two focal points, longitudinal spherical aberration (LSA).

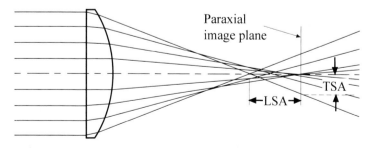

Figure 4.8 Spherical aberration can be measured as the difference between the paraxial and marginal foci (LSA) and the difference between the marginal ray in the paraxial image plane and the optic axis (TSA).

Another measure is the difference between the marginal ray in the paraxial image plane and the optic axis, or transverse spherical aberration (TSA).

Coma is another third-order error. This error occurs for off-axis points on an object. A bundle of rays about a chief ray through the lens will focus at one point in the image plane. A cylinder of rays parallel to the first bundle but at some distance from the chief ray come to a focus at a different location in the plane, and even these rays do not all cross at the same point. The result of this error is a point on the image plane that is smeared into a comet-shaped pattern. In Fig. 4.9, the marginal rays in the meridional plane cross above the chief ray. The difference shown by the arrows is a measure of the coma present.

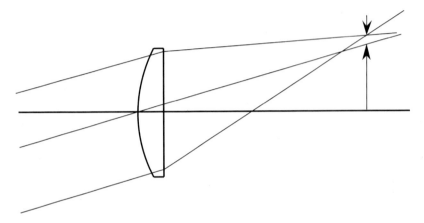

Figure 4.9 Coma in a lens occurs when marginal rays cross at different points in the image plane than the chief ray.

Field curvature occurs because the distance of an object point from the lens increases with the object's height. The simplest argument can be based on a thin-lens approximation. As the object's distance is increased, its image distance decreases. Therefore a planar object in front of an uncorrected lens will be imaged onto a curved surface, called the *Petzval surface* (Fig. 4.10).

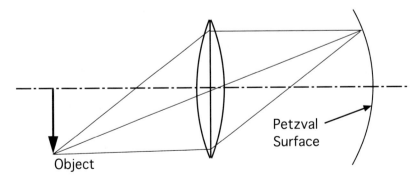

Figure 4.10 Field curvature is an error of image location. A planar object will be imaged onto a curved Petzval surface.

4.1.5 Ray intercept curves

One of the most useful means of assessing the amount and type of error present in an optical system is the ray intercept curve. Although they were invented to allow evaluation of a lens with a small amount of data in an era when computation was long and tedious, ray intercept curves are still useful because they provide easily recognizable information on the types and sizes of aberrations that are present.

On the left side of Fig. 4.11 is a trace of an on-axis parallel ray bundle through a plano-convex lens. The transmitted rays show spherical aberration similar to that shown in Fig. 4.7. A ray intercept curve is a plot of the ray error Δy between the central ray through the lens and the location of that ray in the evaluation plane as a function of the location y of that ray in the entrance place of the lens. The S-shaped curve on the right side of Fig. 4.11 is characteristic of spherical aberration. The amount of bend in the "S" is a measure of the spherical aberration present in the system. It can be shown that the curve will remain the same, but will be rotated about the origin when the evaluation plane is moved along the optical axis.

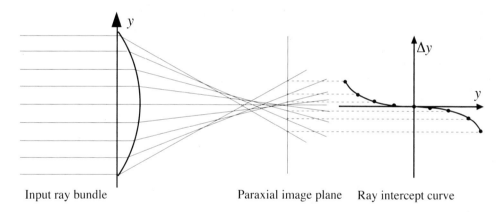

Input ray bundle   Paraxial image plane   Ray intercept curve

Figure 4.11 Ray intercept curve (right) of a lens with spherical aberration. The errors in the rays in the paraxial image plane are plotted as a function of the location of the input ray on the lens.

The ray intercept curve for coma manifests itself as a "U" shape, as shown in Fig. 4.12. In this case the ray errors are measured from the chief ray location in the evaluation plane.

Finally, chromatic aberration is displayed in a ray intercept curve as a series of similarly shaped curves, each plotted for a number of different wavelengths. Figure 4.13 shows ray intercept curves for a lens with spherical and chromatic aberration. The amount of separation between the curves of different wavelengths is a measure of the chromatic aberration present.

This has been a rapid introduction to the basics of lens design. We now look at how these techniques and tools are used for lens design with diffractive optics.

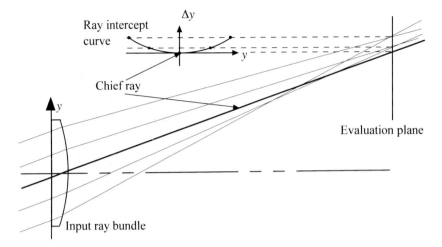

Figure 4.12 Lens with coma and ray intercept curve (top). The errors in the rays in the paraxial image plane are plotted as a function of the location of the input ray on the lens.

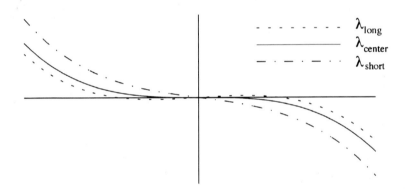

Figure 4.13 Ray intercept curve of a lens with both spherical and chromatic aberration.

4.2 Diffractive Optics Lens Design

The diffractive lens was described from a wave picture in Chapter 1. Essentially, the basis of the diffractive lens is the Fresnel phase plate. In this section we describe the approach to designing a single-surface diffractive lens and look at its limitations.

As an engineer or scientist begins to design a diffractive optical element, he or she has some specific objective in mind. Some of the considerations that go into establishing the required performance are

1. Efficiency, determined by the type of profile that can be generated — binary, multilevel, continuous (kinoform)
2. Optimization of the required wavefront to reduce aberrations and/or increase efficiency — chromatic aberration and thermal sensitivity
3. Fabrication capability available — limitations and tolerances.

4.2.1 The diffractive lens

Consider the thin lens, not as a ray bender, but as a wavefront converter. A plane wave will slow down once it enters a plano-convex lens, as shown in the following discussion. As the plane wavefront traverses the lens, the lens imposes a local phase difference upon the wave. The result will be a converging spherical wave that will collapse at the focal point of the lens, as was shown in Fig. 1.17.

The DOE comparable to this thin lens is a Fresnel phase plate. The bending of light in this case arises from diffraction of the wavefronts at each of the Fresnel zones. By making a phase plate with multiple phase steps, the efficiency of the plate will approach 100%, as described in Sec. 2.4.2 on the blazed grating. Provided the zones are shaped properly, each zone will contribute to focusing the light to the appropriate point.

To see this more clearly, we need to calculate the transition points for each of the zones that are an integral number of wavelengths farther from the focal point of the DOE than the on-axis distance. From Fig. 4.14 we can determine the radius r_p of the pth zone using the Pythagorean theorem:

$$r_p^2 + f^2 = (f + p\lambda)^2. \tag{4.12}$$

Multiplying out and canceling,

$$r_p^2 = 2fp\lambda + p^2\lambda^2.$$

In most cases λ is small, so $p\lambda \ll f$ and the transition points are given by

$$r_p^2 = 2fp\lambda. \tag{4.13}$$

Therefore, to make a diffractive lens of focal length f operating at wavelength λ, a surface should be constructed that changes phase by 2π at those radii. A 2π phase change is equivalent to one wave λ (of optical path difference).

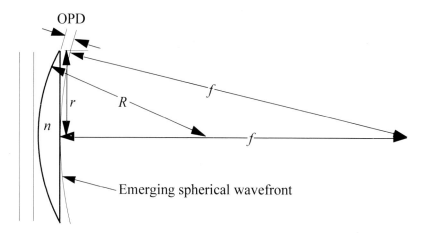

Figure 4.14 Geometry used to determine the locations of the transition points for a diffractive lens (OPD, optical path difference).

The distance between zones, which determines the periodicity of the local diffraction grating at a distance r_p, can be determined from

$$r_p^2 = 2fp\lambda \quad \text{for the } p\text{th zone and}$$

$$r_{p+1}^2 = 2f(p+1)\lambda \quad \text{for the } p + 1\text{st zone.}$$

Subtracting $r_{p+1}^2 - r_p^2 = 2f\lambda$ and expanding the difference of squares:

$$(r_{p+1} + r_p)(r_{p+1} - r_p) = 2f\lambda.$$

For large p, $r_{p+1} + r_p \approx 2r_p$ and $\Delta r = r_{p+1} - r_p$, we obtain

$$2r_p\Delta r = 2f\lambda. \tag{4.14}$$

Think of Δr as $\Lambda(r)$, the local grating period at a distance r_p from the axis. We can then write the local grating period for a diffractive lens as

$$\Lambda(r) = f\lambda/r. \tag{4.15}$$

Note: In the derivation of Eq. (4.13), the phase difference between zones was equal to 2π. In other instances, the phase difference between zones is some fraction of 2π. For example, for the Fresnel zones described in Sec. 1.4, the phase difference is π (the OPD is $\lambda/2$). In that case, the transition points are given by

$$r_p^2 = fp\lambda. \tag{4.16}$$

4.2.2 The phase profile

The phase difference created by the lens is the difference between the plane wave at each point and the phase delay that is due to travel through the lens. At any distance from the optical axis of the lens, the phase difference traversing a lens of axial thickness t is

$$\Delta\phi = 2\pi n t/\lambda - 2\pi t/\lambda = 2\pi(n-1)t/\lambda. \tag{4.17}$$

The thickness of the lens in Fig. 4.14 at any distance r from the axis is given by the "sag" equation:

$$t = r^2/2R.$$

Therefore the resulting phase profile caused by the lens is

$$\Delta\phi(r) = 2\pi(n-1)r^2/2R\lambda. \tag{4.18}$$

It is this profile that will concern us in diffractive optics. The OPD at various points on the wavefront is given by

$$\text{OPD} = \Delta\phi(r)\lambda/2\pi = (n-1)r^2/2R. \tag{4.19}$$

The focal length of the lens can be determined from Fig. 4.14 using the Pythagorean theorem:

$$(\text{OPD} + f)^2 = r^2 + f^2 \rightarrow \text{OPD} = r^2/2f. \tag{4.20}$$

Substituting for the OPD in the previous equation:

$$2f(n-1)r^2/2R = r^2 \rightarrow 1/f = (n-1)/R. \tag{4.21}$$

This is the same equation for the focal length of a plano-convex lens expressed by the lensmaker's equation [Eq. (4.2)].

4.2.3 Generating a single-element design

There are a number of tools that can be used to compute the locations of transitions in a phase profile of a lens. Some can produce binary masks. Although most of these tasks can be done with a user-generated program, there are programs that can relieve much of the tedium that goes with such efforts. Lens design programs such as CODE V,[1] OSLO,[2] and ZEMAX[3] will compute the coefficients of an arbitrary phase profile for specific target values. Some of the programs use symmetrical phase profiles (even powers of the radius). All programs will optimize a diffractive lens or a combination of a diffractive surface on a conventional lens by minimizing aberrations. In some cases, the program will fracture the phase profile and provide mask data for production. Computational programs such as Mathematica[4] can also compute phase transitions and generate masks for simple single elements. Nonconventional optical elements such as beam fanouts, deflectors, and general transforms require different computations. Their computation is described in Chapter 5.

4.2.4 Design of a kinoform lens

Any optical element, assuming no attenuation or reflection, can be considered as a general wavefront transformer that takes an input wavefront and transforms it into a desired output by changing the phase profile of the wavefront. The transmission function of a phase-only optical element can be written as $t(r) = e^{i\phi(r)}$, where r is the radial distance from the center of the component. In the case of a lens, the phase function is approximately given by

$$\phi(r) = (2\pi/\lambda_0)(n_0 - 1)r^2/2R, \tag{4.22(a)}$$

or in terms of the OPD:

$$OPD = (n_0 - 1)r^2/2R. \qquad\qquad [4.22(b)]$$

In the more general case,

$$\phi(r) = (2\pi/\lambda_0)(n_0 - 1)z(r), \qquad\qquad (4.23)$$

where $z(r)$ is the surface profile at the design wavelength. For a kinoform, the profile is fractured at a number of zones to produce the diffractive surfaces. Usually this fracture occurs at any point where the phase changes by 2π and the OPD changes by one wavelength, as we described in Sec. 4.2.1. However, the depth at the fracture can be any integral number of wavelengths (OPD $= m\lambda_0$) and it will still operate as a diffractive component. This is equivalent to using the grating in a higher order. Such profiles are sometimes called *superzones*. A detailed discussion of this structure and its applications is given in Sec. 10.2.4.

4.2.5 A simplification

Once the element has been specified, the phase profile needs to be determined. As we have noted earlier, there is an extremely useful simplification that can be invoked when analyzing the function and performance of diffractive optics. One can consider the profile $\phi(r)$ as consisting of two independent parts: the etch depth and the location of the transition points.

The etch depth determines the operating wavelength of the element. It is equal to

$$d = (2^N - 1)\lambda/[2^N(n - 1)], \qquad\qquad (4.24)$$

where N is the number of masks and n is the refractive index of the substrate on which the profile is fabricated. Each mask etches an additional $\lambda/[2^N(n-1)]$ from the profile. In the limit of a kinoform, the number of masks N goes to infinity and the overall depth of the etch approaches

$$d = \lambda/(n - 1) - \lambda/[2^N(n - 1)] \rightarrow \lambda/(n - 1). \qquad\qquad (4.25)$$

The other part of the profile consists of the locations of the transition points [e.g., Eqs. (4.13) and (4.15)] for each change of the phase profile by 2π or a fraction thereof. The location and the geometry of the transition points determine the function of the diffractive element, be it lens, grating, or general wavefront converter.

The critical element of the design of any DOE is the search for the most efficient and/or easiest-to-fabricate phase profile. Once found, a program or routine can generate the correct transition points for the mask or the tool angle and transition points for a diamond point-turned surface, or the correct electron-beam property or laser dosage to properly expose the photoresist. Next, we need to examine the factors that affect the efficiencies of diffractive optical elements. Although we concentrate on diffractive lenses, many of the conclusions also apply to other elements.

4.3 Efficiency of Multilevel Diffractive Lenses

It is difficult to fabricate smooth profiles for components intended to work with visible light. As pointed out in Chapter 1, multilevel approximations can be generated through the application of binary masks to approximate the ideal profile (Fig. 1.21). For example, in Fig. 4.15, a four-level (two-mask) approximation of a diffractive lens is shown. Note that the profile looks like a blazed grating if the lens is plotted as a function of the square of the radial distance r. This is a Fresnel phase lens with an additional subdivision to the initial 0 and π-phase zones that were illustrated earlier. The additional transitions are at the $\pi/4$ and $3\pi/4$ phase transition points. The locations of the transitions between the 0 and π-phase zones can be written as $r_p^2 = p\lambda_0 f$. For the N-level mask, the radii will be $r_p^2 = 2p\lambda_0 f/2N$.

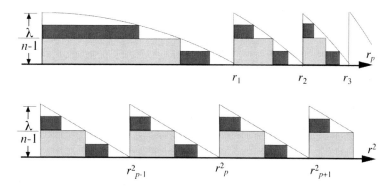

Figure 4.15 Four-level approximation to the smooth profile (kinoform) of a diffractive lens with 100% efficiency. The upper plot is the structure itself and the lower plot is the structure as a function of the radius squared, showing the periodic nature of a lens in r^2 space.

The efficiency of the lenses increases as the number of levels is increased, and the multilevel profile becomes a better approximation of the kinoform. It was shown for the case of the blazed grating in Fig. 2.10 and for other components such as lenses that the efficiency of the optics increases in a nonlinear manner as the number of levels increases. Using Eq. (2.34), Table 4.1 lists the efficiency as a function of levels. The "linear" approximation to the profile consists of linear segments between the 0 and 2π-phase levels of each transition in place of the steplike segments of the binary mask profiles.

4.3.1 At other wavelengths

Diffractive lenses are designed to operate at a specific wavelength. When the lens is used at another wavelength, the focal length will change and its efficiency will

Table 4.1 Efficiency of diffractive elements as a function of number of levels.

Levels	Efficiency (%)
2	40
4	81
8	95
16	99
Linear	99
Polynomial	100

be reduced. A measure of the mismatch in the design is given by the wavelength detuning parameter, α:

$$\alpha = \frac{\lambda_0}{\lambda} \frac{n(\lambda) - 1}{n(\lambda_0) - 1}. \tag{4.26}$$

In the case of a lens, it can be shown that the focal length changes to $f' = (\lambda_0/\lambda)f/m$, where f is the focal length at the design wavelength λ_0 and m is the diffraction order in which the component is operating (usually $m = 1$).

In Fig. 4.15, the profile of a diffractive lens as a function of r^2 is shown to be a periodic ramp. Using this profile, it can be shown that the efficiency of a diffractive lens kinoform is given by

$$\eta = \frac{\sin^2[\pi(\alpha - m)]}{[\pi(\alpha - m)]^2}, \tag{4.27}$$

where α is the detuning parameter and m is the diffraction order. Plots of the efficiency of the kinoform lens as a function of wavelength for several orders are given in Fig. 4.16. Note that at the design wavelength all of the light is diffracted into the

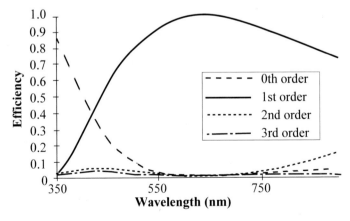

Figure 4.16 Efficiency of a diffractive lens as a function of wavelength for a design wavelength of 633 nm for various diffraction orders.

first order and the lens maintains its efficiency in a broad range of wavelengths about the design wavelength.

This variation of efficiency with wavelength can be interpreted as either a wavelength error or as an error in the etch depth. An etch depth d' that differs from the correct value of d can be interpreted as an element whose design wavelength is $\lambda' = d'(n-1)$ as given in Eq. (4.25). Thus, errors in etch depth can be calculated as departures from the design wavelength. If the errors in depth are small, the changes in refractive index as a function of wavelength are also small and α [Eq. (4.26)] can be written as

$$\alpha = \lambda_0/\lambda = d/d'. \tag{4.28}$$

If the departure from the design depth d_0 is εd_0, then $d = d_0 \pm \varepsilon d_0$ and $\alpha = d_0/(d_0 \pm \varepsilon d_0)$ or $\alpha = 1/(1 \pm \varepsilon) = 1 \pm \varepsilon$ for small ε. When α is inserted into the equation for the efficiency and p is set equal to one for the first order, the efficiency is approximately equal to

$$\eta \approx \sin^2(\pi\varepsilon)/(\pi\varepsilon)^2. \tag{4.29}$$

As one can see from Table 4.2, the error in depth has to be greater than 5% before it has a substantial effect on the efficiency of a diffractive element. However, as described later, small errors in the locations of the transitions can result in drastic reductions in the efficiency of a component.

Table 4.2 Efficiency of a diffractive component as a fractional error in the etch depth.

Error	Efficiency (%)
0	100.0
0.01	100.0
0.02	99.9
0.05	99.2
0.1	96.8
0.2	87.5

4.3.2 Efficiency of diffractive lenses

One question that usually arises once the fascination of diffractive components has subsided is: How good are these things? With all of the sharp edges in this element, how well can it perform? One way of analyzing this is to treat the diffractive lens as a grating. If the lens were perfectly formed, in effect, if the grating was correctly blazed for the design wavelength, all of the light should be diffracted in the first order. There is a small region at the transitions where there can be "shadowing," an effect that was depicted in Fig. 2.11, or scattering that is due to the structure of the walls at the transitions. This light may be directed into other orders or scattered into image space.

The effect of a collimated beam of light incident on a diffractive lens is shown in Fig. 4.17(a). The edges of the light field for the three lowest orders of the lens are also shown in the figure. The light in the first order is directed to the focal point, while the second-order contribution focuses at half the focal distance and then diverges to evenly illuminate the focal plane. The undiffracted (zero-order) light continues on as if there were no lens there. In the focal plane of the diffractive lens, almost all of the light is found in the first-order peak [Fig. 4.17(b)], but the zero-order and higher-order contributions manifest themselves as a weak, broad background. This has little effect on the image quality, but it does affect the image contrast. This shows up as a precipitous drop in the contrast at zero-spatial frequencies for the modulation transfer function [Fig. 4.17(c)], a standard method of characterizing the performance of lenses.

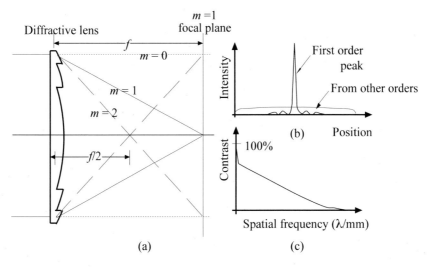

Figure 4.17 Distribution of light from a diffractive lens. (a) Direction of light from the three strongest orders. (b) Intensity in the focal plane of the lens. Other orders contribute to the background. (c) Modulation transfer function of the lens.[5]

4.3.3 A diffractive optics lens and its limitations

One of the problems with the phase surface we have just described is that the phase function changes so rapidly at the edge of the lens that the periodicity becomes very small. Depending on the fabrication technique, it may or not be possible to make masks that contain features to the edge of the lens.

The transition radii for a kinoform (λ steps) are $r_p^2 = 2fp\lambda$. For a level-one mask ($\lambda/2$ steps), the locations of the transitions for the binary mask are given by

$$r_p^2 = 2fp\lambda/2 = fp\lambda. \tag{4.30}$$

For the Nth-level mask, the radii will be

$$r_p^2 = 2fp\lambda/2^N = fp\lambda/2^{N-1}.\tag{4.31}$$

Using the same approach to determine the local periodicity of the diffractive lens that we used in Sec. 4.2.1, it can be shown that the difference between radii at the r_p radius is

$$\Delta r_p = f\lambda/2^N r_p.\tag{4.32}$$

If our fabrication technique will only permit a minimum feature size of Δr_{min}, this limits the f-number of the lens that can be fabricated:

$$\Delta r_{min} = f\lambda/2^N r_{max} = f\lambda/2^{N-1} D_{max} = 2F\lambda/2^N,\tag{4.33}$$

where $F = f/D_{max}$ is the f-number of the lens. Thus the smallest f-number lens that can be fabricated is given by

$$F = 2^N \Delta r_{min}/2\lambda.\tag{4.34}$$

For example, if the minimum feature size is 10 μm, the minimum f-number that can be achieved for a first-level mask operating at a helium-neon laser wavelength is

$$F = 10 \ \mu m/0.6328 \ \mu m = 15.8.$$

For higher-level masks, the f-number doubles with each higher level. If the feature size is 1 μm, the minimum f-number will be 1.5 for a first-order mask and $f/12$ for a third-order (eight-level) DOE.

This discussion is based on the approximation we made an Eq. (4.13) (i.e., $p\lambda \ll f$). However, if one has submicron fabrication capability, the value of p can be very large and the approximation no longer holds. It can be shown that a more exact minimum-zone spacing in a diffractive lens of a given f-number is

$$\Delta r_{min} = \lambda\left[1 + (2F)^2\right]^{1/2}.$$

4.4 Hybrid Lenses

A diffractive surface can be combined with a classic optical element, such as a lens, to produce a hybrid component. In this section we examine the addition of diffractive profiles to conventional lenses to correct aberrations. In Chapter 10, we will extend this approach to multielement lenses and to systems in which diffractive surfaces can be used to provide thermal compensation. We begin, however, with a very simple approach to hybrid optics.[6]

4.4.1 Correcting chromatic aberration with diffractive surfaces

It can be easily shown, using the approach to color correction described in Sec. 4.1.3.1, that the dispersion of a diffractive optics surface is incredibly large compared with even the strongest flint glass. Only a small amount of diffractive power is required to correct a chromatic lens.

Using Eq. (4.15) for the local grating period, $\Lambda(r) = f\lambda/r = \lambda/\phi r$ and solving for the power of the lens,

$$\phi_c = \lambda_c/r\Lambda(r). \tag{4.35}$$

The product $\Lambda(r)r$ represents the profile and therefore is a constant of the lens. We see that the power of the lens changes linearly with wavelength. Therefore, if the lens is designed for a specific wavelength, say λ_c as given in Eq. (4.35), then the power at any other wavelength $\phi(\lambda) = \lambda/r\Lambda(r)$ can be determined by inserting the construction term $r\Lambda(r) = \lambda_c/\phi_c$. The power of a diffractive lens at other than the design wavelength is given by

$$\phi(\lambda) = (\lambda/\lambda_c)\phi_c. \tag{4.36}$$

Now consider the powers of the lens at the center, long, and short wavelengths:

$$\phi_c = \lambda_c/r\Lambda(r); \; \phi_l = \lambda_l/r\Lambda(r); \; \phi_s = \lambda_s/r\Lambda(r) \tag{4.37}$$

and calculate the amount of chromatic aberration as

$$\Delta\phi = \phi_s - \phi_l = (\lambda_s - \lambda_l)/\Lambda(r)r. \tag{4.38}$$

So the equivalent V-number for a diffractive lens is a ratio of wavelengths:

$$V = \phi/\Delta\phi = \lambda_c/(\lambda_s - \lambda_l), \tag{4.39}$$

which is always negative.

Not only is a diffractive surface dispersion negative, it is extremely strong! For example, in the visible region: $V = 587.6/(486.1 - 656.3) = -587.6/170.2 = -3.45$. If you compare this value with flint glasses that have V-numbers in the range of 20 to 50, you can see that it would take only a modest amount of diffractive power to correct a refractive lens (Fig. 4.18). Note that the V-number for a diffractive surface has nothing to do with its shape. It is dependent only on the wavelengths, so the dispersion is the same for all diffractive surfaces over the same wavelength band.

Given a 100-mm effective focal length (EFL) lens made of BK7, a standard crown glass, it is easy to determine how much diffractive power is required to correct this lens at the C and F lines using the approach described in Sec. 4.1.3.1. The V-number of BK7 is 64.2 and the power of the lens is 1/100 mm or 0.01 mm^{-1}. The chromatic correction relation was given as

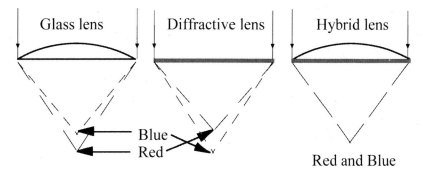

Figure 4.18 Color correction of a hybrid lens. A glass lens (left) with positive dispersion focuses blue light inside the focus for red light, whereas a positive-power diffractive lens (center) has negative dispersion. When combined as a hybrid lens, the two dispersions cancel, creating an achromatic lens.

$$\phi_a/V_a = -\phi_b/V_b. \tag{4.11}$$

Inserting the values for ϕ_a and V_a for the glass lens and V_b for the diffractive surface, the optical power of the diffractive lens ϕ_b is found to be

$$\phi_b = -0.01 \text{ mm}^{-1} \times -3.45/64.2 = 5.374 \times 10^{-4} \text{ mm}^{-1}.$$

So the power of the diffractive lens is a small fraction of the refractive lens and its focal length is $f_b = 1.861$ m. The total power of the lens is now $\phi = \phi_a + \phi_b = 0.01 + 0.0005374 = 0.01054 \text{ mm}^{-1}$ or $f = 94.5$ mm.

4.4.2 Starting with a singlet

Although it is possible to compute the effect of adding a diffractive surface to a classic design using a number of analytical techniques and tools, the easiest way is to use one of several optical design programs. We first show the type of data needed for a plano-convex lens and the type of output that is produced. We then show a diffractive lens that does the same thing. Finally, we combine them to make a hybrid lens.

For this example, we use a plano-convex lens from the Newport catalog (SPX022). This is a 100-mm EFL, $f/5$ (20-mm diam.) quartz lens. The prescription is

Radius (mm)	Thickness (mm)	Aperture (radius, mm)	Material (index)
45.702	4.79	12.7	Quartz ($n = 1.458464$)
0 (flat)*	BFL	12.7	Air (1.0)

*It is a convention in lens design that a flat surface, which has an infinite radius of curvature, can be specified as zero.

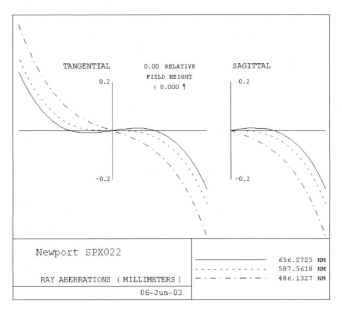

TANGENTIAL   0.00 RELATIVE   SAGITTAL
                    FIELD HEIGHT
         0.2   0.2
                    (0.000)

         -0.2   -0.2

Newport SPX022

                                    ─────────────── 656.2725 NM
                                    --------------- 587.5618 NM
RAY ABERRATIONS (MILLIMETERS)   ─·─·─·─·─·─·─· 486.1327 NM
                        06-Jun-03

Figure 4.19 Aberration curves for a stock plano-convex lens showing a large amount of spherical aberration and transverse color.

The specifications are object distance, infinity; field angles, 0 deg; wavelengths (nm) 656.3, 589.7, and 485.1.

We are only evaluating one field point (on the optical axis), so rays are parallel to the axis, making the specifications somewhat simpler than normal. The performance of this lens is shown by its ray intercept curve in Fig. 4.19. The only aberrations evident for a system that is evaluated on-axis are spherical aberration and chromatic aberration. As we showed earlier (Fig. 4.10), the S-shaped curve indicates spherical aberration and the y coordinate gives the amount. The separations between the curves for the various wavelengths indicate the amount of chromatic aberration of the lens.

4.4.3 Example: diffractive surface on a quartz window

Optical design programs are based on tracing rays through optical systems according to the laws of geometrical optics. It would seem that such programs could not account for the effects of a diffractive optics surface since the optical design programs are ray based, whereas diffractive surfaces are wave-based creatures. However, lens designers are very ingenious souls and they have found ways to adapt standard ray-trace routines to simulate diffractive surfaces within a conventional optical design program. [One notational problem arises. In ray-based systems, ϕ stands for the power of the lens and in wave-based systems it stands for the phase of the wave $\phi(r)$. To distinguish the two quantities, we use boldface type for the lens power $\boldsymbol{\phi}$ in this section.]

As a simple example, we start with a quartz window (two parallel flat surfaces) and add a diffractive profile to one of the surfaces. For example, for an $f/5$ 100-mm

EFL plano-convex lens, the power ϕ of the lens is given by:

$$\phi = 0.01 \text{ mm}^{-1} = (n-1)(0 - 1/R).$$

From Eq. [4.22(b)] in Sec. 4.2.4, OPD $= (n-1)r^2/2R$. Inserting $R = -(n-1)/\phi$, [Eq. (4.4) with $f = 1/\phi$], the profile for this lens can be specified as

$$\text{OPD} = -r^2\phi/2. \tag{4.40}$$

Any diffractive lens can be optimized by adding additional terms to the OPD:

$$\text{OPD}_{\text{def}} = A_1 r^2 + A_2 r^4 + A_3 r^6 + A_4 r^8. \tag{4.41}$$

When the phase constants A_1 through A_4 are allowed to vary during optimization, an optimum surface is generated. Then this function is fractured by integral wavelengths to find the transition locations for the kinoform. The OPD function and the phase function are related by

$$\phi(r) = 2\pi/\lambda \text{OPD} = 2\pi/\lambda(A_1 r^2 + A_2 r^4 + A_3 r^6 + A_4 r^8). \tag{4.42}$$

This function can be fractured to generate a phase mask.

However, not all design programs use the same phase function to describe the diffractive surface. For example, the manual for CODE V, a product of Optical Research Associates, says that the "(e)valuation of the polynomial gives the OPD in lens units at the construction wavelength HWL." In comparison, the program OSLO Pro from Lambda Research defines the phase function as

$$\phi(r) = 2\pi \mathbf{dor}/\mathbf{dwv}(\mathbf{df1}r^2 + \mathbf{df2}r^4 + \mathbf{df3}r^6 + \mathbf{df4}r^8 + \cdots),$$

where $\mathbf{dor}$ is the diffraction order, $\mathbf{dwv}$ is the construction wavelength, and $\mathbf{df1}\ldots\mathbf{df4}$ are the phase coefficients. The diffraction wavelength is entered in microns in the program, but the phase factor is evaluated using the wavelength in lens units. This is a distinction not found in any OSLO documentation. ZEMAX (Focus Software) uses a phase function but normalizes the radius "to scale the x- and y-intercept coordinates of the rays. This is done so that the coefficients of the polynomial all have units of radians." When any of the various programs are executed, the component looks considerably different from the usual line of lenses, as is shown in Fig. 4.20.

In the lowest approximation of the OPD, OPD $= A_1 r^2$, so $A_1 = -\phi/2$ using Eq. (4.39). For our example, $A_1 = -0.005$. When A_1 is multiplied by $2\pi/\lambda$ and λ is 633 nm, the first term in the phase function is $\pi/\lambda f = -0.004963$. Because the function is expressed as a polynomial of even powers, the additional degrees of freedom permit the program to optimize the phase function by reducing spherical aberration. The optimized values in the three programs of the first four coefficients of the surface to produce a lens with the characteristics in the example are given in Table 4.3.

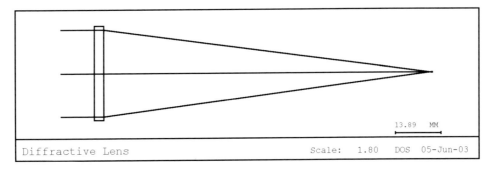

13.89   MM

Diffractive Lens   Scale: 1.80   DOS 05-Jun-03

Figure 4.20 A lens with a diffractive surface on one side of a quartz window. The coefficients of the phase surface that produces this focusing element are given in the text.

Table 4.3 Phase coefficient for a pure diffractive lens for three commercial optical design programs*.

Program	Function	A_1	A_2	A_3	A_4
CODE V	OPD	-5.00×10^{-3}	-5.7512×10^{-28}	-4.0541×10^{-30}	-3.7945×10^{-32}
ZEMAX	Phase	-4.963×10^{-3}	$+1.223 \times 10^{-7}$	$+2.44 \times 10^{-11}$	-1.496×10^{-13}
OSLO Pro	Phase	-4.963×10^{-3}	-2.864×10^{-23}	$+6.414 \times 10^{-10}$	4.379×10^{-30}

*The different constants for the two phase functions are probably due to the different merit functions used by the respective programs to optimize the lens performance. During the seven or so years since the routines were first run, the values given by "same" program have differed substantially, except for A_1.

4.4.4 Combining refractive and diffractive surfaces

The example given here is not much of a hybrid lens since it consists of a diffractive surface on a flat quartz window. A more instructive case consists of a plano-convex lens with a curved refractive surface on one side of the lens and a diffractive structure on the "plano" surface. The introduction of a diffractive surface permits the designer to do more than just correct chromatic aberration. One can also correct for spherical aberrations.

When we add a diffractive surface to the plano surface of the lens, it is necessary to specify a surface that is described by its phase variation using either the OPD or phase function, as discussed early. To optimize the lens and reduce the aberrations, we then permit the radius of curvature of the first surface to vary along with the coefficients (A_1, A_2, A_3, and A_4) of the diffractive surface. The final values of the refractive lens are similar to the earlier example:

Radius (mm)	Thickness (mm)	Aperture (radius, mm)	Material (index)
48.80292	4.79	12.7	Fused silica ($n = 1.458$)
0 (flat)*	BFL	10	Air (1.0)

The coefficients of the diffractive polynomial on the second surface, computed in Code V are

A_1	A_2	A_3	A_4
-3.1365×10^{-4}	$+3.5180 \times 10^{-7}$	-2.4772×10^{-11}	$+4.4228 \times 10^{-14}$

The performance is given in the ray intercept curves shown in Fig. 4.21. Note that the scale of this curve is the same as that in Fig. 4.18 and that the lens is well corrected compared with its nondiffractive counterpart.

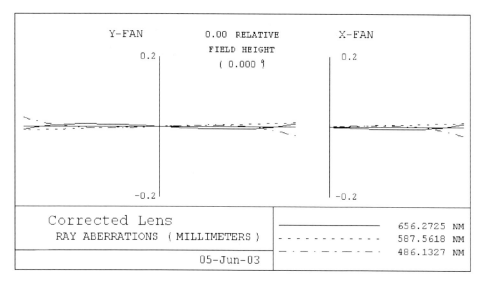

Figure 4.21 Correction of spherical aberration and color using a diffractive surface. Ray intercept curves are shown for a plano-convex lens with a diffractive surface on the plano surface.

The central wavelength shows almost no spherical aberration at all. The ray intercept curve snakes back and forth across the *x*-axis. The long and short wavelengths show small amounts of higher order spherical aberration and excellent color correction since the curves cross one another. Considering that there are only two surfaces to manipulate, the addition of the diffractive surface to this lens shows the effectiveness of diffractive optics.

In some lens design programs there are macros that permit one to convert the diffractive surface data to a series of binary masks. Sometimes this may be helpful. But in some cases it is just as easy to take the coefficients and enter them into a separate program that generates the mask. Some authors have used a graphical calculation routine written in Mathematica[7] to generate such masks. In Chapter 10, we will extend the approach we have just described to multielement lenses and thermal compensation of lenses.

References

1. Optical Research Associates, Pasadena, CA.
2. Lambda Research Associates, Rochester, NY.
3. Focus Software, Tucson, AZ.
4. Wolfram Research, Champaign, IL.
5. D.A. Buralli and G.M. Morris, "Effects of diffraction efficiency on the modulation transfer function of diffractive lenses," *Appl. Opt.* **31**, p. 4389 (1992).
6. T.W. Stone and N. George, "Hybrid diffractive-refractive lenses and achromats," *Appl. Opt.* **27**, p. 2960 (1988).
7. D.C. O'Shea, "Generation of mask patterns for diffractive optical elements using Mathematica," *Computers in Physics* **10**, pp. 391–399 (1996).

Chapter 5

Design of Diffraction Gratings

5.1 Introduction

The discussion in Chapter 4 centered on the design of diffractive optics using conventional lens design techniques, in which each ray incident on the DOE is mapped to one output ray. The refractive lenses and mirrors and the diffractive elements described in Chapter 4 manipulate the wavefront by a 1-to-1 mapping [Fig. 5.1(a)]. Diffractive structures can also be used by the optical engineer for wavefront splitting, in which elements are designed to send light into multiple diffraction orders [Fig. 5.1(b)]. In this chapter we discuss the characteristics and design techniques for diffractive optics that perform a 1-to-N mapping. The techniques used to design complex gratings can be applied to more general classes of diffractive optics, but for consistency we illustrate these design approaches using gratings.

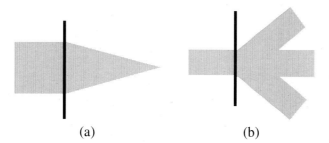

(a)	(b)

Figure 5.1 Wavefront mapping. (a) 1-to-1 wavefront mapping—the diffractive element mimics a refractive lens. (b) 1-to-3 mapping—the diffractive element acts as a beam-splitting grating.

5.1.1 Splitting a wavefront

When a beam of light passes through any periodic structure, it will be diffracted into multiple orders. The period of the repeated structure determines the angular separation between the orders. A small period creates large angular separation, while a large period results in closely spaced output beams. The angle θ_m of each output beam of order m relative to the zero-order beam is given by the grating equation:

$$\sin\theta_m = m\lambda/\Lambda, \tag{5.1}$$

where Λ is the period of the structure.

Although the grating equation determines *where* the light is directed, it does not determine the relative power directed into each of the diffracted orders. This power distribution is dictated by the shape and nature of the surface profile within a single grating period. As we saw in Chapter 2, if the shape within a unit cell of the periodic structure is an ideal blaze, all of the diffracted light is transmitted into the +1 order. An identical shape with a smaller phase depth will split light between the 0 and +1 orders, as well as diffracting smaller quantities into the other orders. The more complicated grating structures allow the wavefront to be split into essentially arbitrary power distributions between the output orders. A wide variety of phase structures, as well as attenuating or blocking features, can be incorporated into the unit cell of a periodic structure. The ability to harness this design freedom is a major benefit of diffractive optics. The creation of these special gratings and splitters requires an understanding of the design variables and, in all but the simplest of cases, a computer.

5.1.2 A 1 × 3 grating

A simple periodic structure, the binary phase grating, was introduced in Chapters 2 and 3. In Fig. 2.9, the binary phase grating was shown as a first approximation of a continuous-blazed grating. The binary approximation, however, is only 40.5% efficient in the +1 diffraction order. This binary profile is an equally good approximation of an ideal blaze in the opposite direction, so it follows that this grating is also 40.5% efficient in the −1 diffraction order. Thus, a linear binary grating with a fixed period can also be considered to be an 81% efficient beamsplitter or 1×2 grating (see Fig. 2.8). The remaining light is diffracted into higher orders.

To provide additional insight into the control of the various grating orders, we introduce an additional degree of freedom into our binary phase grating design: the duty cycle, which is the ratio of the feature dimension b to the period of the grating Λ. For the present, we will consider only binary gratings. Also, we will retain the same etch depth d that provides a half-wave optical path difference between the two grating levels.

To understand the binary phase grating, it is best to analyze it in the context of Fourier optics using the convolution theorem. From the physical definition of the grating, one can quickly create a mathematical description of the element using the convolution operation and the Dirac delta function, as discussed in Chapter 2. The mathematical description of the unit cell is two offset rect functions with a phase constant applied to the second one (Fig. 5.2). The offset is accomplished with the delta function.

$$\text{unit cell} = f(x) = \left[\text{rect}(x/b) \otimes \delta(x + b/2)\right]$$

$$+ \left\{\text{rect}[x/(\Lambda - b)] \otimes \delta[x - (\Lambda - b)/2]\right\}e^{i\phi}. \tag{5.2}$$

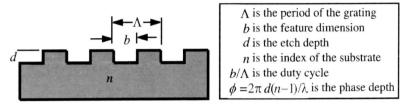

Figure 5.2 The physical definitions of a binary phase grating.

Having defined the unit cell, an infinite grating can be generated with one additional convolution with the comb function:

$$\text{grating} = f(x) \otimes \text{comb}(x/\Lambda). \qquad (5.3)$$

As discussed in Sec. 2.3, the discrete orders propagate at angles θ_m given by the grating equation [Eqs. (2.26) and (5.1)]. Therefore the Fourier transform of the grating is only defined at discrete intervals, the orders of the grating, which correspond to the transform of $\text{comb}(x/\Lambda)$. The result of this Fourier transform is also a comb function: $\text{comb}(\Lambda q)$, which has a unit value at $q = m/\Lambda$, where m is any integer.

The Fourier transform of the unit cell, $\Im[f(x)] = F(q)$, forms an envelope function, the function that multiplies the comb function and gives the amplitude and relative phase of each diffracted order. As noted in Sec. 2.4.1, the Fourier transform of the $\text{rect}(x)$ function is $\text{sinc}(q)$. Therefore, the transform of $f(x)$, which describes the unit cell, is the sum of two sinc functions with offset phase terms.

$$
\begin{aligned}
F(q) = {} & \frac{b}{\Lambda} \frac{\sin(\pi b q)}{\pi b q} \exp(i2\pi q b/2) \\
& + \frac{(\Lambda - b)}{\Lambda} \frac{\sin[\pi(\Lambda - b)q]}{\pi(\Lambda - b)q} \exp[i2\pi q(\Lambda - b)/2].
\end{aligned} \qquad (5.4)
$$

In the case of our earlier binary phase grating, where $b = \Lambda/2$ and $\phi = \pi$, this simplifies to

$$F(q) = \frac{1}{2} \frac{\sin(\pi b q)}{\pi b q} \Big[\exp(i\pi b q) - \exp(-i\pi b q)\Big]. \qquad (5.5)$$

Restating this envelope function for the values where the comb function is nonzero, the output of the grating is given by

$$|F(q)|_{q=\frac{m}{\Lambda}} = \frac{\sin(m\pi/2)}{m\pi/2}[i\sin(m\pi/2)], \qquad (5.6)$$

which equals zero for all even values of m. For odd values of m, the diffraction efficiency from the grating is given by

$$\eta_m = \left| f\left(\frac{m}{\Lambda}\right) \right|^2 = \left(\frac{2}{m\pi}\right)^2, \qquad (5.7)$$

as listed in the table in Fig. 2.8.

As the grating depth or duty cycle changes, the grating output is still defined by the envelope function in Eq. (5.4). For example, by varying the duty cycle b/Λ, the intensity of the zero, first, and second orders of the grating can be plotted, as shown in Fig. 5.3. We see that away from the 50% duty cycle, the zero order now contains energy. At a duty cycle of 26.5%, the first and zero orders cross each other. This means that the zero and the first orders all have the same efficiency, in this case 22%. So a binary grating with this duty cycle serves as a 1×3 beamsplitter. The efficiency of this beamsplitter, calculated as that for the 1×2 splitter, is 66%, because 22% of the incident beam is directed at each of the three orders ($m = -1, 0, +1$).

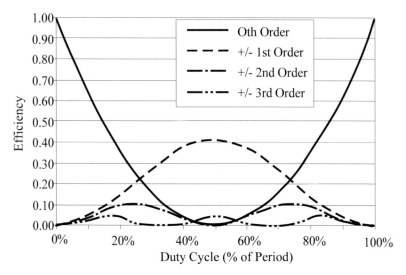

Figure 5.3 Efficiency of a binary phase grating of π-phase depth as a function of grating duty cycle for different diffraction orders. Note that at a duty cycle of 26.5%, the zero order and first orders have the same efficiency, creating a 1×3 grating.

Another way to control the output pattern of a binary element is through its phase depth. Instead of changing the duty cycle, we begin with the 50% duty cycle of the 1×2 grating and vary the phase depth between 0 and 2π (Fig. 5.4). The 1×2 grating occurs when the depth equals π. Note that just as in the case of varying the duty cycle, varying the phase depth produces a 1×3 grating with a 50% duty cycle when the phase is equal to 0.64π. However, the efficiency for this grating is somewhat greater than the previous example. Each of the orders contains about 28.8% of the incident beam, giving a grating efficiency of 86%. Therefore,

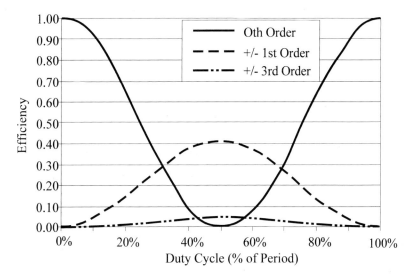

Figure 5.4 Efficiency of a binary phase grating with a 50% duty cycle as a function of grating depth. At a phase depth of 0.64π, the first order and zero orders have the same efficiency, creating a 1 × 3 grating. Because of the 50% duty cycle, the second (and all even orders) are zero.

although most binary diffractive structures are based on a π-phase depth, there are other choices. This is a relatively isolated case. The use of phase depth as a design variable is usually only advisable for simple binary grating designs with a limited number of other variables.

Although the 1 × 3 grating is a simple example, it illustrates a number of key points for diffractive elements. Virtually every diffractive element diffracts light into multiple orders. For any diffractive grating, the angular spacing of the orders is determined by the local grating period. The relative power in the diffraction orders is determined by the structure within the unit cell. In review, this fact is evident in all the examples used to this point in the text.

The blazed gratings introduced in Chapter 2 are assumed to have structure within each period to direct light into the +1 order, as are the diffractive lenses described in Chapter 4. The structure in the unit cell can be changed to diffract light into different orders without changing the direction of propagation for each of those orders. In the case of the 1 × 3 grating, the grating period is set to define the desired diffraction angle for the +1 and −1 orders, and the unit cell structure is optimized to split light equally between these orders and the zero order.

5.1.3 Complex fanouts

As the complexity of the desired output pattern increases beyond one or two diffraction orders, the calculations needed to determine the unit cell very quickly become difficult. The unit cell of a grating becomes extremely elaborate because more orders must be controlled and the output pattern becomes more complex (Fig. 5.5). Dammann and others have developed a straightforward method of cal-

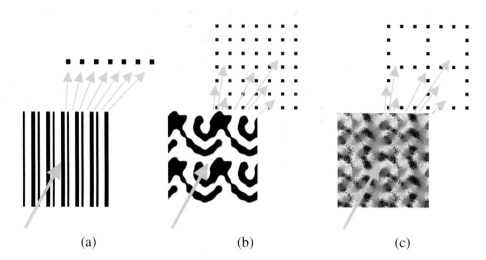

Figure 5.5 Fanout gratings can range from simple to complex. (a) The design of a one-dimensional binary phase grating with a few orders can be calculated directly. (b) Two-dimensional gratings with many orders, and (c) multilevel phase gratings require iterative computer algorithms.

culating the unit cell structure for $1 \times N$ gratings.[1–3] The technique assumes a binary structure with a phase depth of π, and casts the problem as a set of N non-linear equations and N unknowns. As N increases, the mathematics can become daunting. Also, this technique does not easily extend to two-dimensional gratings or to multilevel diffractive structures. However, because this technique was one of the first proposed for designing fanout gratings, they are often called *Dammann gratings*.

The industrial requirements for fanout gratings vary in complexity from 1×3 elements used in many compact disk heads to illumination systems requiring two-dimensional beam fanouts with as many as 270,000 diffracted orders. For the majority of these applications, designers rely on iterative computer algorithms to calculate the structure of the grating unit cell. Given a desired far-field output, there are a number of proven techniques for calculating the grating structure that will generate the required output, as described in the rest of this chapter.

5.2 Design Approaches

The approach to the design of a diffractive grating depends on its function. The designer must consider the element's role in the overall system, the wavelength of operation, and the relationship of performance to cost. Although a grating is defined by the far-field intensity pattern it generates, it is affected by the environment in which it is used. Aberrations from other components in the system, position errors, and beam characteristics can change the desired output and must be considered as parts of the overall grating design problem.

For example, if a grating is placed before a focusing lens that is diffraction limited on its axis but has severe coma, then the zeroth order beam will be a diffraction-limited spot in the focal plane, but the higher orders will be blurred. Although gratings are usually designed for one wavelength, they are often used in systems with a finite waveband of operation. At the other wavelengths, the grating will perform differently. For example, the diffraction angles of the grating orders change with wavelength [Eq. (5.1)], as does the power in the zeroth diffraction order [Eq. (5.4)]. Also, some applications require high diffraction efficiency and tight power balance between the diffracted orders, while others do not. These requirements help to determine the necessary fabrication precision, the number of phase levels, the optical coatings, and the substrate quality. They can dramatically affect the manufacturing requirements for a grating, including cost. Some designers have discovered that their very elegant grating designs cannot be fabricated within their budget.

The fabrication and physics of the device must be considered before the design process begins. Although the design approaches we discuss can be used for a wide variety of gratings, we limit our discussion to diffractive gratings fabricated using photolithography and etching. Because commercially available photomasks are limited to roughly 1 μm of lateral resolution, and are typically limited in overall size to between 100 and 150 mm, they present limitations on the performance of the DOEs that can be fabricated.

Let us assume that our application has a 10-mm incident beam size with a wavelength of 1 μm. We can make the following statements: (1) The grating can have diffractive features that are one wavelength in lateral size. (2) The beam can illuminate up to 100,000,000 features (10,000 1-μm features on a side) as it passes through the element. The first statement suggests that scalar design techniques could be inaccurate for the application. Because the size of the diffractive features approaches the wavelength of light, discontinuities in the electric field at the phase steps must be analyzed, as illustrated in Chapter 3, using vector theory techniques to provide an accurate simulation of the diffractive performance.

The second statement can be recast in terms of a space-bandwidth product (SBWP). This term, derived from communication theory, is the product of the frequency of a transmission times the duration of its signal. It defines the upper limit at which information can be transmitted. In the spatial domain there is a simple analogue to this. Here the space-bandwidth product is a measure of the information that can be encoded on a beam of light by a diffractive element. It is defined as

$$\text{SBWP} = \frac{L \times L}{\Delta \times \Delta}, \tag{5.8}$$

where L is the beam dimension and Δ is the minimum resolvable feature. So, in the example given here, $\text{SBWP} = 10^8$.

A common design technique is to define a unit cell as being an array of discrete "pixels" of size Δ by Δ. The design problem is then solved by varying the etch depth of each pixel until an optimum is found. In the extreme, the unit cell could be as large as the beam, in which case the number of design variables is equal to the space-bandwidth product. In designing such a diffractive element, the phase value for each resolvable "pixel" in the element would have to be set independently of the others. The thousands of independent variables provide a tremendous absolute design freedom, but they also present a massive computational problem. Strategies are needed to reduce the number of variables to an efficient workable level without affecting the resulting design. This requires a thorough understanding of the diffractive grating and its desired function.

The designer also needs to consider the complexity of the solution before choosing a design approach. In the next section, different algorithms for producing an efficient design are presented. Each is best suited to a particular class of problems. Fortunately, there are many proven and well-documented options. For simple designs, one can employ Dammann's method as referenced earlier, or direct inversion. For more complicated gratings, deterministic algorithms that converge rapidly can be used, even for large numbers of variables. If the desired output is very complex, stochastic optimization techniques such as simulated annealing or genetic algorithms may be required. All these approaches are discussed in Sec. 5.5. Although the examples shown will use scalar diffraction theory, it is important to remember that these iterative techniques can be applied to both scalar and vector simulation.

5.3 Design Variables

To begin the design of a diffractive grating, we must mathematically describe or "encode" the unit cell of the grating. Any technique we use involves some assumptions. At this stage it is important that the assumptions be based upon well-understood factors, such as the fabrication method. For a photolithographically generated grating, the unit cell structure is constructed from discrete, uniformly etched levels. Two general encoding techniques are used:

1. A binary technique assumes two phase levels and varies feature sizes by changing the transition points. One of the phases is assigned to each of the defined areas.
2. An array technique uses a rectilinear array of pixels within the unit cell.

Each pixel is assigned one of a number of preset phase depths. Encoding a diffractive structure by assigning the phase depths based upon some rule, then optimizing the transition locations, is a logical extension of diffractive lens design. It is the approach used for creating Dammann gratings as well. The designer assumes phase levels of zero and π, then treats the transition locations in x and y as design variables, as shown in Fig. 5.6(a).

This encoding approach can be coupled with an iterative technique in which the transition locations are changed methodically and the performance of the grating is evaluated after each change. Another approach for a 2D element is to depart from the strictly rectangular geometry of Fig. 5.6(a) and include patterns made up of trapezoids [Fig. 5.6(b)].[4–6] Because the phase transitions are not rectilinear,

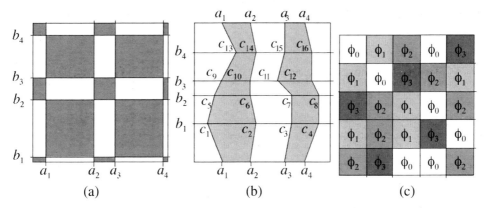

Figure 5.6 Phase information can be encoded in a grating unit cell with 25 features in a number of ways. Phase values can be set and the transition locations treated as design variables, or the transitions can be set and the phase values varied. (a) A Dammann grating unit cell has 0 and π phase values, and the transition locations are computed to achieve the desired output pattern. (b) Trapezoid-based encoding permits more variation in the location of phase transitions. (c) The most common encoding technique has fixed pixels, with the phase values of each pixel as a design variable; for a four-level grating, the pixel can take on one of four preset phase values.

a trapezoid-based encoding adds substantial design freedom. The unit cell of 25 phase elements shown in Fig. 5.6(a) is a function of eight variables; the trapezoid-based unit cell has the freedom of 16 additional variables. One reason trapezoids are used is that the software used to create photomasks for the electronics industry accepts both rectangular and trapezoidal shapes.

The second technique, using a rectilinear array of pixels within the unit cell, is the most common. The unit cell consists of a fixed array of uniformly sized pixels. Then the phase value in each pixel is treated as a variable [Fig. 5.6(c)].[7,8] This approach has many advantages. It works equally well for binary and multi-level diffractive structures. Because the pixels are rectangular (often square), the pattern readily translates to mask data. Also, the pixels are uniformly spaced, so it is straightforward to calculate the Fourier transform of the unit cell on a computer using a discrete Fourier transform (DFT) or a fast Fourier transform. For the purpose of illustrating different design algorithms, we will focus on this encoding method. However, the optimal encoding techniques and design algorithms will vary with the application. All of the design algorithms discussed in Sec. 5.5 can be used to optimize gratings encoded by any of these techniques. For the majority of design problems, the phase depths are assumed to have fixed values. In some limited cases, such as the 1×3 grating illustrated in Fig. 5.4, the grating depth can also be treated as a design variable.

The absolute design freedom for diffractive gratings is tremendous. Photomask technology permits the designer to set the number of design variables as high as the space-bandwidth product. However, this large number of independent variables is usually impractical. The designer has to make smart choices that limit the unit cell to an appropriate size; then an encoding technique must be chosen that balances adequate design freedom against a reasonable number of variables. Encoding techniques that use too much design freedom can slow the design process, while those with too little design freedom are over-constrained and cannot produce a good design. The choice of the appropriate coding is helped by experience.

5.4 Direct Inversion

Given a desired grating output pattern, how do you calculate the grating function that produces it? Because of the complexity and number of variables, a computer is the tool of choice for grating design. A straightforward approach is direct inversion. The first step is to map the desired pattern onto a grid of discrete spots. This grid corresponds to the diffraction orders of the grating (see Fig. 5.7). The discrete spots will be located at angles θ_m, as defined in Eq. (5.1). Based on that equation, the grating period can be related to the target spot pattern. Once the period is known, the problem reduces to defining the phase structure within the unit cell.

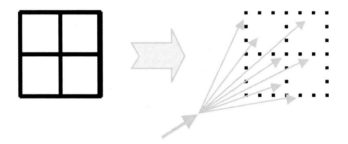

Figure 5.7 The desired output pattern is mapped onto a grid of discrete spots. The grating period can be derived from the angular separation of these spots.

Once the grating period is determined, the unit cell can then be divided into a two-dimensional array of pixels. Two factors determine the number of pixels used to encode the unit cell: (1) The number of pixels must be large enough to produce the required performance. (2) The maximum number of pixels is limited by the minimum pixel dimension, which must be greater than or equal to the minimum resolvable feature, which is in turn dictated by the fabrication process. Traditionally, it was desirable to have the number of pixels be an even power of 2 for ease of calculation, but with mixed-Radix algorithms, that is no longer a requirement.

For example, if the angular separation between spots is 0.5 deg at $\lambda = 633$ nm, the grating period will be 72.5 μm from the grating equation. The minimum resolvable feature for standard photomasks is about 1 μm, and 64×64 is an excellent computation grid for FFTs. Therefore, we can define our unit cell as 64×64 with a pixel dimension of 1.133 μm. The design problem is now reduced to finding an array of phase values for which the intensity of the Fourier transform closely matches the target spot pattern. Because we are using FFTs, we define the unit cell to be in the coordinate domain and the target output pattern to be in the spatial frequency domain.

One means of solving this problem is direct inversion. The output intensity pattern is the square of the field in the spatial frequency domain. Therefore we know the amplitude to be the square root of the intensity. This neglects the phase component of the field. Because the relative phases of the target spots are usually not important for the performance of the grating, we make an arbitrary assumption: when the target intensity pattern is converted to complex amplitude values, a random phase value will be assigned. The target complex field is centered on a 64×64 array of output spots. Mathematically, the spot pattern can be expressed as

$$I_{m,n} = |F_{m,n}|^2, \tag{5.9}$$

where

$$F_{m,n} = A_{m,n} \exp(i\phi_{m,n}). \tag{5.10}$$

The amplitude $A_{m,n}$ is equal to zero for all locations where the output spot pattern is not defined, and the phase $\phi_{m,n}$ is a randomly assigned number. Values of m and n range from -31 to $+32$. Once the output is encoded, an inverse Fourier transform of $F_{m,n}$ will produce the desired grating unit cell. Unfortunately, the inverse transform, $f_{m,n}$, varies continuously in both amplitude and phase. To encode this in the unit cell, we set the amplitude of each pixel in the field to unity. The phase of each pixel is truncated to one of the $N = 4$ preset phase values. This sequence is depicted in Fig. 5.8. The inverse transform

$$f_{m,n} = a_{m,n} \exp(i\varphi_{m,n}) \tag{5.11}$$

is truncated to

$$f'_{m,n} = (1.0) \exp(i\varphi'_{m,n}), \tag{5.12}$$

where

$$\varphi'_{m,n} = 0, 2\pi/N, \ldots, 2(N-1)\pi/N, \text{ where } N = 4 \text{ for this example.} \tag{5.13}$$

Although this technique is straightforward, it has some drawbacks that become evident when the grating is "played back" by performing a Fourier transform on the assigned amplitude and phase. Ideally, the output spot pattern should look like the target. However, the truncation introduces substantial noise and pattern nonuniformity, which is unacceptable for many applications. Close examination of the playback transform in Fig. 5.8 shows that the nonuniformity of intensity in the target spots is very large. In some places the target spots are indistinguishable from the noise.

Direct inversion is often used to design elements for optical information processing, such as matched spatial filters for pattern recognition. For this application, both the intensity and phase of the reconstruction are critical, so the "random" phase values described here have to be replaced with target phase information. Often in these cases the efficiency of the element is secondary to the signal-to-noise ratio for the device. In these cases, a layer of metal may be evaporated onto the

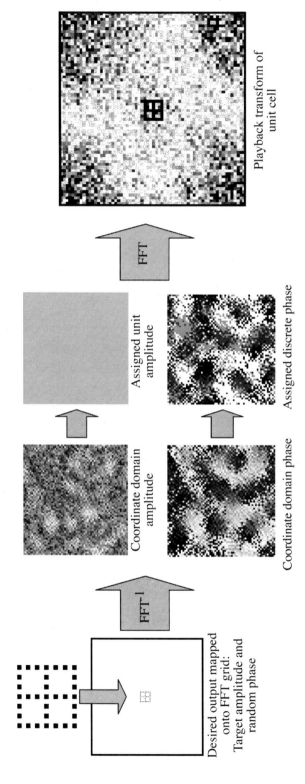

Figure 5.8 Direct inversion. The target output intensity is mapped onto an FFT grid and random phase values are imposed. The field is inverse transformed, then truncated to unity amplitude and four discrete phase values. Grating performance is calculated with another FFT to simulate reconstruction.

completed diffractive structure and patterned to produce amplitude modulation. The amplitude modulation introduced by this metal layer is binary, effectively setting the transmittance of a pixel to zero, unless the metal patterning process can resolve apertures within each pixel. If subapertures can be achieved, elements can be created with finite phase and amplitude modulation. Although this sacrifices efficiency, it gives superior reconstruction fidelity.

For elements that do not employ amplitude modulation, reconstructions from diffractive optics designed with direct inversion can be very noisy. This noise is the result of mapping the coordinate domain amplitude to unity. In Fig. 5.8 we see that changing the amplitude to unity is a dramatic truncation in which a great deal of information is lost.

The changes in the coordinate domain complex field that are due to phase truncation are inversely proportional to the number of phase levels in the diffractive element. With a large number of phase levels, little information is lost. Simulations with and without amplitude truncation confirm the logical assessment that the playback nonuniformity is largely due to the amplitude being set to unity. Earlier simulations with varying levels of phase truncation (i.e., 2, 4, 8, or 16 phase levels), show good fidelity for values of 8 or higher, but very noisy playback for 2-phase encoding. This makes sense because more severe truncation will sacrifice more information in exchange for simplicity in fabrication.

The purposes of applying a random phase value to each of the target amplitudes in $F_{m,n}$ is that it helps distribute the power in the near field. If the phase in the frequency domain were uniform across the field, ($\phi_{m,n}$ = constant), the amplitude after inverse transform would be very high near the center of the field and nearly zero elsewhere. This would result in even poorer performance after amplitude truncation. The random-phase term applied in the frequency domain better distributes the amplitude values in the coordinate domain. There is, however, still significant modulation (Fig. 5.8).

The reconstruction would be much better if the inverse transform of the target field were nearly uniform in amplitude because less information would be lost through amplitude truncation. There is a set of variables by which we can control the amplitudes in the coordinate domain field. The amplitude values in the target field are fixed, but by varying the phase values ($\phi_{m,n}$) in the target field, it is reasonable to expect that some set of $\phi_{m,n}$ will minimize the amplitude modulation in the coordinate domain field. This frames a critical question: How can one determine which set of phase values, applied to the initial target field, will produce minimal amplitude modulation upon transform, and therefore superior performance upon playback? Arriving at that set of target phase values can be used as the objective of several iterative optimization techniques.

5.5 Iterative Design

The Fourier transform relationship between the grating unit cell and its far-field output pattern was previously described. In the case of a binary grating with only two transitions, the unit cell was expressed by Eq. (5.2) and its Fourier transform, the output intensity by Eq. (5.6). A discrete or fast Fourier transform defines the field on a sampled grid, with amplitude and phase values at each grid point. If the amplitude and phase information are known in one domain, the values can be calculated in the other. For a relatively small grid of 64×64 points, the unit cell is defined by 4096 amplitude values and 4096 phase values. Each of the 4096 amplitude and 4096 phase values in the far field can be calculated if all the values in the unit cell are defined. Gerchberg and Saxton[9] showed that this relationship could be extended. If the amplitude values are known in both domains, but the phase values are not, one still has 8192 known values and 8192 unknown values. The transform relationship can be used to generate a set of 8192 equations and 8192 unknowns. Through this mathematical inversion, the phase values can be retrieved when the amplitude is defined in both domains. Similarly, the amplitude values can be calculated if the phase information is known in both domains. This illustrates both the versatility and complexity of the Fourier transform relationships that we use in grating design. A relatively simple design problem has thousands of variables. More sophisticated design problems may have several million variables. Arriving at the optimum solution is a formidable task.

Mathematicians have studied the optimization of multivariable systems. Most useful techniques fall into three general categories: bidirectional algorithms, unidirectional algorithms with deterministic constraints, and unidirectional algorithms with stochastic constraints.[10] Bidirectional algorithms are ideal for simple systems with a large number of variables. Unidirectional algorithms are preferred when the system operation cannot be easily inverted. For unidirectional algorithms, deterministic constraints will cause the system to converge rapidly to some optimum. Stochastic constraints allow the design algorithm to evaluate several possible solutions before arriving at an optimum. Although all three categories are used for grating design, the choice of which to use is determined by the complexity and function of the element.

5.5.1 Bidirectional algorithms

In the direct inversion example, the Fourier transform relationship was used because it is easy to perform an inverse transform. If the optical system function can be easily inverted, the designer has the option of utilizing a bidirectional algorithm. Optical system functions that can be inverted include Fourier transforms, Fresnel transforms, and certain other linear operations. Gratings are almost always analyzed using a Fourier transform.

In a bidirectional algorithm (Fig. 5.9), a first "estimate" at the grating cell is generated, typically using a random number generator to assign starting values. The pixel values for the estimate are then truncated, based on encoding constraints such as cell transitions or quantized phase depths. A system transform is applied, and performance constraints are imposed on the field in the output plane. Some performance evaluation is made at this point. The performance constraints are usually intensity requirements. An inverse system transform is then applied, and the

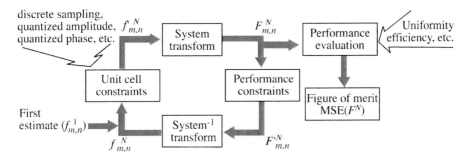

Figure 5.9 A bidirectional algorithm for optimizing diffractive optical elements (MSE, mean-square-error).

resulting field is treated as an updated estimate of the unit cell. The cycle is repeated and the system converges to some optimum estimate of the grating unit cell. During each cycle, a "figure of merit" is calculated in a manner similar to the process described for optimizing lenses in Chapter 4. It is a measure of the difference between the current values and the target values. As in lens design, the figure of merit is used to determine the performance of the current design and the rate of progress toward the desired target pattern.

In a Fourier transform system, if the amplitude values are known in both domains, the phase values can be calculated. As noted earlier, optimization of a 64×64 grid reduces to a set of 8192 equations and 8192 unknowns. The calculations become very involved. The solution for the unknown phase values can also be determined using a bidirectional algorithm in a fraction of the time required to invert the 8192 equations. If the coordinate domain field amplitude and target domain field amplitude are applied as the constraints, Gerchberg and Saxton[9,11] showed that the coordinate and target field phase information can be reconstructed, usually in just a few iterations. This method of phase retrieval can be used for grating design as well. For a case where the *actual* near- and far-field amplitudes of a Fourier transform system are known, the Gerchberg–Saxton algorithm will converge to define the unknown phase fields. When designing a grating, one has the desired unit cell characteristics and the output plane amplitude. A similar bidirectional algorithm can be used to predict the phase values that will most nearly fit the constraints.[12] Even with an inexpensive computer, these calculations can be performed in seconds.

The bidirectional algorithm to perform grating design can be expressed mathematically. As discussed in the direct inversion case, the objective is to identify a set of target phase values ($\phi_{m,n}$). If the optimum phase values are used in the target field, the amplitude upon inversion to the coordinate domain will be nearly uniform, and the coordinate domain phase information can be encoded as a phase grating. Encoding the phase grating from a field with uniform amplitude means that minimum information will be lost when the amplitude in the coordinate domain is set to unity. Because the effects of amplitude truncation are minimized, the grating performance will be much better. This is best described in the context of

Fig. 5.9. A random first estimate, f^1, is entered, and the unit cell constraints of unit amplitude and quantized phase are applied.

$$f_{m,n}^1 = a_{m,n}^1 \exp(i\varphi_{m,n}^1) \rightarrow \text{constraints} \rightarrow f_{m,n}^{\prime 1} = (1.0)\exp(i\varphi_{m,n}^{\prime 1}). \quad (5.14)$$

An FFT is performed:

$$F_{m,n}^1 = A_{m,n}^1 \exp(i\phi_{m,n}^1) = FFT[f_{m,n}^{\prime 1}]. \quad (5.15)$$

The output field is evaluated, and the intensity constraint is applied, resetting the amplitude values throughout the field to the target amplitude values, but the phase information, $\phi_{m,n}$, is preserved:

$$F_{m,n}^1 \rightarrow \text{constraints} \rightarrow F_{m,n}^{\prime 1} = A_{m,n}^{\prime} \exp(i\phi_{m,n}^1), \quad (5.16)$$

where $A_{m,n}^{\prime}$ is the target amplitude of the object at coordinate m,n. An inverse FFT is now performed, and the resulting field is now the new estimate, f^2.

$$f_{m,n}^2 = FFT\left(F_{m,n}^{\prime 1}\right). \quad (5.17)$$

As this loop is repeated, the unconstrained phase values are rapidly driven to values that minimize the noise and the nonuniformity exhibited in the direct inversion example. The results of our design example are shown in Fig. 5.10. The progress of the optimization is determined by a merit function, which is commonly defined as the mean-square-error (MSE) of the playback intensity with respect to the target intensity. This is not the only possible merit function. Some merit functions place

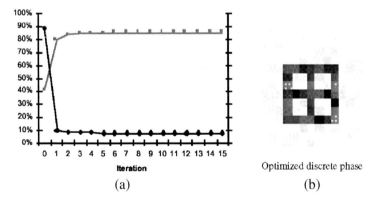

(a)

Optimized discrete phase

(b)

Figure 5.10 Bidirectional optimization. (a) Plot of the merit function and efficiency as a function of iteration number, showing that the solution converges rapidly, reaching a minimum MSE within twelve iterations. (b) The playback transform of the design is significantly better than the direct inversion design example, but still shows some nonuniformity. The MSE is 0.87, which is significantly better than the MSE of 8.2 for direct inversion.

heavier weight on the absolute efficiency of the design, while others measure the intensity uniformity in the target orders. For this example, the merit function measures the MSE of the reconstruction. This is based on the deviation of the value of each point in the output field from its target value. As the quality of the reconstruction improves, the value of the MSE merit function decreases:

$$\text{MSE}(F^1) = \frac{1}{MN} \sum_{m,n} \left(\left| F^1_{m,n} \right| - A'_{m,n} \right)^2. \tag{5.18}$$

A system with N variables can be thought of as having an N-dimensional solution space. The 64×64 design example has a mind-numbing 4096-dimension solution space. The bidirectional algorithm converged, within twelve iterations, to a "best" value for the merit function in this complex solution space. But, did it really arrive at the best solution? The efficiency is very high (85%, better than direct inversion at 46%), but the uniformity of the spot pattern could be better. The best solution can be inferred from the figure of merit, in this case the mean-squared-error is 0.87. So how do we determine that we have achieved the "best" solution? This is a critical question for all iterative optimization algorithms, and the process we use affects our probability of success. To better explain the optimization process, it is helpful to use a simple analogy: the blind prospector.

Consider the plight of a blind prospector looking for a seam of ore. From past experience he knows that the geology of the region is such that the richest veins are found in the deepest valleys. Since he cannot see the terrain, he must make his way to the lowest point by feel. The problem is that he has no idea if the valley he enters contains the lowest point in the region or is just one of many local minima. Where he ends up is entirely dependent on where he starts.

Consider the simplest case with only two variables. The merit function can be graphed as a function of these variables (see Fig. 5.11). The resulting contours for the merit function will show high and low values. We can think of the surfaces of the merit function as contoured surfaces similar to a small section of a mountain range. The optimum design is represented by that point where the merit function is lowest: the deepest valley. A bidirectional design algorithm will evaluate one point in that section of the "mountains" and, in response to the constraints, will move

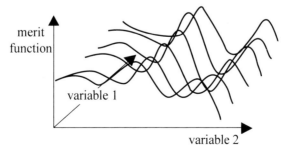

Figure 5.11 The design optimization of a system with N variables can be visualized as searching for the deepest valley in an N-dimensional mountain range.

downhill, like our blind prospector. A bidirectional algorithm moves downhill very effectively, finding the lowest merit function in a few iterations. If the starting point is within the same valley in solution space, the algorithm will converge to the same solution. If the starting point is in another valley, it will converge to the lowest point in *that* valley. This algorithm and many like it are deterministic. As with the prospector, the starting point of the search determines the solution it finds. The quality of the solution is (sometimes unwittingly) set by the first estimate. See the box on optimized beam shaping on page 111. It should be pointed out that the techniques of direct inversion, bidirectional algorithms, and unidirectional algorithms can be applied to broader design tasks than grating optimization. One example is given at the end of this chapter.

Like the blind prospector, the bidirectional algorithm can only move downhill. Therefore, it cannot guarantee that there is not another valley that is deeper. This is the pitfall of deterministic algorithms. If the blind prospector is placed on gently rolling hills with one broad valley, he can easily find his way down to the truly best solution, the global minimum. But if the design is very sensitive to small changes in some key variables, the merit function in solution space is very craggy. The prospector is much more likely to get stuck in a little nook, and never find the mother lode. These little nooks in solution space are local minima. If the solution space is complex, it can limit the effectiveness of the algorithm. If there are many local minima, the bidirectional algorithm has a much higher likelihood of getting stuck.

In our 64×64 design example, we began with a random guess for the first estimate and proceeded to the best design within that "valley" in the solution space. This is now a search within a "mountain range" projected into a 4096-dimension space. There are doubtlessly a large number of local minima in this space. The algorithm converged rapidly to one local minimum. A different first estimate will most likely yield a different solution that has a chance of having a lower MSE than the first solution. How will you know when you have found the best solution to the design problem? You won't. But, for many applications, the results of a bidirectional algorithm are completely adequate.

A bidirectional algorithm has many advantages for grating optimization. Besides being computationally simple, it converges rapidly, even for very large numbers of variables, and produces a much better result than direct inversion. It does not guarantee the optimum design, because of the deterministic form of the algorithm. Nondeterministic, or *stochastic*, algorithms perform a more general search of the solution space, but require substantially more computer power.

5.5.2 Simulated annealing

A well-characterized stochastic optimization algorithm is *simulated annealing*. Simulated annealing is a unidirectional algorithm. A system transform is calculated in the forward direction only, and the merit function is evaluated. The variables are changed, and the impact of the change on the merit function is used as the basis for improving the design.[13–15] This process is illustrated in Fig. 5.12.

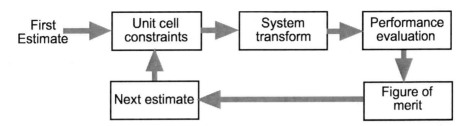

Figure 5.12 A unidirectional algorithm for optimizing diffractive optical elements.

Simulated annealing is loosely patterned after the crystallization of metals. Molten metal crystallizes as it cools. Slow cooling encourages a highly ordered crystal structure, resulting in a stronger material. Rapid cooling results in small, discontinuous crystals with irregular boundaries. This results in very brittle and hard materials. The difference stems from the orientation of the molecules when the metal becomes solid. While liquid, the orientation of each molecule is an independent variable. As a bar of metal freezes, these millions of variables are frozen. If the freezing process is carefully controlled, the molecules orient themselves into larger, stronger crystals. When mathematicians embraced what smiths had long known, the concept of simulated annealing was transformed into a numerical technique for arriving at "highly ordered" arrays of independent variables.

With grating design, the "order" of the independent variables is measured by the merit function. Beginning with a first estimate, the forward system transform is performed repeatedly as the variables that define the unit cell are changed randomly. These unit cell variables are analogous to the molecular orientations in the metal. Initially, the new estimates are accepted arbitrarily. Some changes lower the value of the merit function, others do not. Changes that lower the value of the merit function (which means a better design) are accepted. As the algorithm iterates, some of the changes that *increase* the value of the merit function are also accepted. However, as the optimization proceeds, the probability of accepting these changes is decreased. This is analogous to slowly cooling the metal. The basic simulated annealing algorithm is depicted in Fig. 5.13.

Our 64×64 design example can be optimized using a simulated annealing algorithm. The key step is the addition of a probability variable, t, which is analogous to the temperature of classic annealing. This variable is a measure of the probability of accepting a change that increases the merit function. The variable t is usually related to the iteration number, so that it becomes smaller as the algorithm progresses. The unit cell variables can take on the same phase values used in the inverse transform,

$$\varphi'_{m,n} = 0, 2\pi/N, \ldots, 2(N-1)\pi/N, \text{ where } N = 4 \text{ for this example.} \quad (5.13)$$

Note that N in Eq. (5.13) is 4, the number of phase levels in our grating. It should not be confused with the N-dimensional solution space described earlier.

```
t₁ = 1.0
Initialize first estimate
Evaluate performance
For (k = 1, max) {
Perturb estimate
Evaluate performance
If better
Update estimate
If worse
p = Random (0.0, 1.0)
if (p < tₖ)
Update estimate
t₍ₖ₊₁₎ = t₁/k
}
```

Figure 5.13 Pseudo-code representation of a simple simulated annealing algorithm.

In the first estimate, these values are assigned randomly and the system transform is performed. The merit function (MSE) is the mean-squared difference of the target field intensities from the field produced by transforming the first estimates:

$$F_{m,n}^1 = FFT\left[(1.0)\exp(i\varphi_{m,n}'^1)\right]. \tag{5.19}$$

Next, the unit cell phase values are randomly perturbed to create a next estimate. With each iteration k, the transform and evaluation steps are repeated to generate $F_{m,n}^k$, and a decision is made if:

1. $MSE(F^k) < MSE(F^{k-1})$, then accept the new estimate.
2. $MSE(F^k) > MSE(F^{k-1})$, then generate a random number p. If $p < t_k$, accept the new estimate.

This process is repeated until the maximum number of iterations is reached or an acceptable merit function is achieved. During this process the variable t is slowly lowered, and with it the probability of accepting a change that increases the merit function value. For our design example, this takes appreciably longer than the bidirectional algorithm, but it also achieves a significantly lower mean-square-error (Fig. 5.14). The simulated annealing algorithm requires, in this example, nearly 10 times as many computations to produce a design. The result has a mean-square-error 7 times smaller than the bidirectional algorithm. Although the efficiency of the bidirectional algorithm is higher (85% > 76%), the actual fit to the target function is our true figure of merit in this case. For most applications, the improvement in uniformity would far outweigh the slightly lower efficiency. Thus the simulated annealing solution is superior to the bidirectional result. There is a clear benefit, but at a cost measured in computation time.

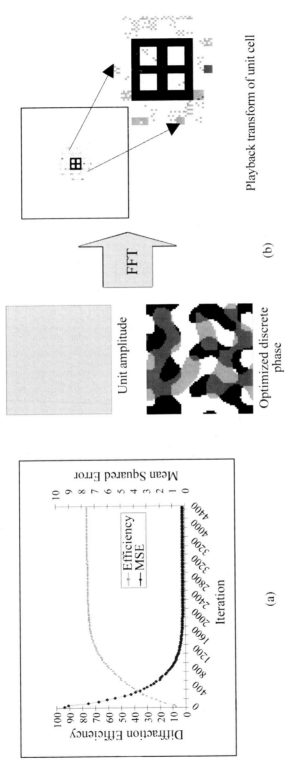

Figure 5.14 Simulated annealing optimization. (a) Plot of the merit function and efficiency as a function of iterations, showing that the solution converges rapidly, reaching a minimum in about 2000 iterations. (b) The output of this design is a better match to the target than deterministic algorithm results. The MSE is 0.12.

To explain the difference between simulated annealing (a stochastic algorithm) and bidirectional optimization (a deterministic one), we return to our blind prospector. To construct an analogy to simulated annealing, we have to equip our prospector with very strong legs. Thus, at the beginning, he can leap great distances about the solution space, trying to find a deep, promising valley. If a leap takes him to a peak instead of a valley, that's okay, he is exploring! As time goes on, he tires of all the leaping about, so if a leap takes him to a higher point in the "mountains," he may decide to backtrack to his last starting point and try again. Soon, he is rejecting more and more of the leaps that take him uphill. His leaps shorten as he finds a deep valley, until he is striding downhill in a manner similar to the bidirectional algorithm. He still may not find the lowest point in the solution space, but the solution he finds is no longer determined by where his first estimate is. Also, with the extended exploring, he has a much lower chance of getting stuck in a little nook.

Our goal is to search the N-dimensional solution space for the true global minimum, without the computational chore of creating a detailed map of the entire space. To create such a map for our simple design example would require 10^{2432} computations, which is impractical. Therefore, most grating design problems reduce to this type of search within the solution space. While the simulated annealing search required less than 10^{10} computations, the bidirectional algorithm required only about 10^6 computations.

Simulated annealing algorithms were initially developed as a technique to search for solutions to systems with many variables. Because the approach is very scalable, it has been used for grating designs with as many as 4,194,304 variables (2048×2048), although a more practical limit for a 500-MHz personal computer is about 512×512. This is adequate for all but the most sophisticated grating designs. Simulated annealing requires substantially more computer power than deterministic algorithms, such as bidirectional optimization, but it is virtually guaranteed to converge to a better solution. Although strict conditions must be met for it to converge to the global minimum, in general, it produces better-performing gratings than direct inversion or deterministic methods. For many applications, this level of performance is very desirable.

A key parameter of simulated annealing is the cooling rate, t. If the system is cooled slowly, the algorithm will have a better chance of finding a really good solution. In fact, for very slow, controlled cooling, there is a mathematical proof that the algorithm will achieve a global minimum: the best solution. Conversely, if the system is cooled rapidly, it becomes little more than a deterministic search. Starting t with a value of 0, the algorithm proceeds to the nearest local minimum. In contrast, if the cooling rate is too slow, the algorithm will take a very long time to produce a superior design. It is important to choose a cooling rate for the variable t so that simulated annealing produces a reasonable solution within a reasonable period of time.

5.5.3 Genetic algorithms

A second type of stochastic optimization algorithm is collectively referred to as a genetic algorithm.[15,16] Just as simulated annealing is based on the physical model of crystallizing metal, genetic algorithms are modeled after natural selection. Because of the relatively slow pace of species procreation, evolution is not usually viewed as an optimization process, but it is a powerful technique nonetheless. Consider the complexity of a typical mammal's DNA. There are about 100,000 genes or variables that describe that animal. The viability of one creature is subject to many issues, but the response of a breeding population to its environment is dramatic. One can visualize that placing a population of dogs in an environment that favors a loud bark, herding ability, and a warm coat produces a collie in less than a hundred generations.

Capturing the power of natural selection in a grating optimization algorithm may appear somewhat contrived, but the results are compelling. The amplitude and phase variables within the unit cell are concatenated together as a "chromosome" (Fig. 5.15). Each estimate is treated as an individual within the population. The optimization is performed on a population of estimates in parallel, in an "environment" that favors better merit functions. The individual estimates are not updated, as in simulated annealing. Each estimate is transformed and its performance evaluated from its merit function. Estimates are then "bred," via a chromosome crossover operation similar to the biological analogue, until a next generation of estimates is created. The selection of parent estimates is based upon their merit functions. Better-performing estimates have a higher probability of passing their performance attributes on to succeeding generations. This manner of selective breeding is repeated until an optimum is found.

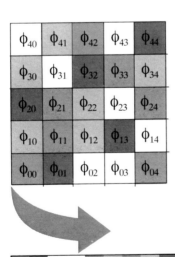

- The phase values of the unit cell are mapped into a linear chromosome.

- Each estimate in the population is represented as a different chromosome.

- Estimates with better merit functions have better chance of becoming parents.

Figure 5.15 Mapping the unit cell into a chromosome.

A genetic algorithm is unidirectional, as described earlier (Fig. 5.12). It operates on a population of estimates in parallel.[17,18] Each estimate undergoes a system transform in the forward direction only. After the merit function is evaluated for all the estimates, crossover operations are used to create the next population. These successive groups of estimates are called *generations*. As generation upon generation evolves a solution to our grating design, the members of a population come to resemble each other and eventually appear identical. As in animal husbandry, selective breeding for a simple set of attributes or a merit function quickly leads to inbreeding. An inbred population (one in which all the estimates are extremely similar) is an indication that the genetic algorithm has found a minimum. This occurs as a matter of course in every optimization. If, however, it occurs too rapidly, the solution may be a local minimum. To limit this possibility, genetic algorithms borrow another concept from natural selection: mutation. The arbitrary reassignment of a few random genes within a population is adequate to temper the rapid inbreeding, and permits the algorithm to escape some local minima (just as simulated annealing can sometimes accept a poorer estimate). Figure 5.16 shows a pseudo-code that describes the steps in a genetic algorithm.

```
Initialize first population of estimates
   (Random)
For (k = 1, max) {
Evaluate performance of estimates
Regenerate Population
Select parents
Crossover
Mutate Population
   }
```

Figure 5.16 Pseudo-code representation of a simple genetic algorithm.

To illustrate the operation of a genetic algorithm, we return to our 64×64 design example. For each generation number, k, there are l members of the population. The allowable values for the phase in each pixel of the unit cell are still set by our fabrication constraint, $N = 4$:

$$\varphi_{m,n} = 0, \pi/2, \pi, 3\pi/2. \tag{5.20}$$

In the starting population of estimates ($k = 1$), these phase values are assigned randomly. The system transform is performed and the merit function (MSE) calculated l times, once for each estimate in the population. The merit functions for the estimates are then ranked as

$$R^{k,l} = \text{MSE}(F^{k,l}). \tag{5.21}$$

The ranking of the estimates within a population is based upon their merit, $\text{MSE}(F^{k,l})$. Once the estimates are ranked, parents are chosen at random, with a probability weighted according to the ranking similar to the one shown in Fig. 5.17.

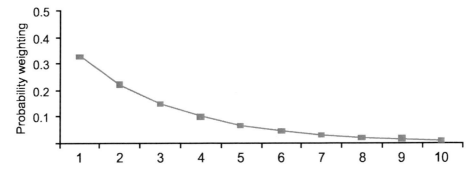

Figure 5.17 Probability weighting of parent selection based on ranking. Better-performing estimates have a higher probability of being selected as parents. In our example, the worst estimate has less than a 1% chance of being selected as a parent, the best has a 33% chance.

When two parents have been selected, their chromosomes are aligned and a crossover operation is performed. The crossover randomly assigns break points in the chromosome string. At each break point, the values in the string are swapped (see Fig. 5.18). Values from the first string are assigned into the second string, and vice versa. The number of breakpoints can be set or chosen at random. Genetic algorithms that use several break points within a string tend to optimize more rapidly, but this advantage disappears as the number of break points gets to about 5 to 10% of the total number of variables in the chromosome. Via the crossover operation, two parent estimates generate two "offspring" estimates for the next generation.

The final operation in a typical genetic algorithm iteration is mutation. After a new generation of estimates is calculated, each with the crossed attributes of its parents, a small number of the phase values within the population are changed randomly. Like natural mutation, the changes are very rare, affecting a tiny fraction of the variables within the population (see Fig. 5.19). Mutation sites are chosen at random within the entire population, and the phase values at those sites are reset with a random number generator, subject to the constraint of Eq. (5.20). When parent estimates are nearly identical, the offspring of those parents are going to appear identical as well. Mutation inserts a critical random element, much as it does in nature. These small changes have a significant impact, reducing the effect

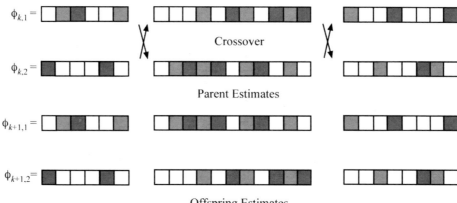

$\phi_{k,1} =$

$\phi_{k,2} =$

Crossover

Parent Estimates

$\phi_{k+1,1} =$

$\phi_{k+1,2} =$

Offspring Estimates

Figure 5.18 The crossover operation produces two new estimates from parent estimates to create a next generation of estimates. Break points are chosen at random. Variable values between break points are swapped. Offspring have some performance attributes of both parent estimates.

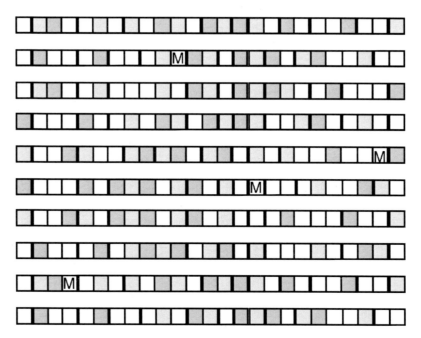

Figure 5.19 Mutation inserts a critical element by randomly changing the value of a few variables within the population. This operation acts to prevent isolation within a local minimum.

of inbreeding and preventing the genetic algorithm from getting caught in a local minimum.

Using genetic algorithms, our 64×64 pixel design example required about 800 generations to arrive at a minimum (see Fig. 5.20). The convergence curve looks very similar to that for simulated annealing. Although the actual grating looks

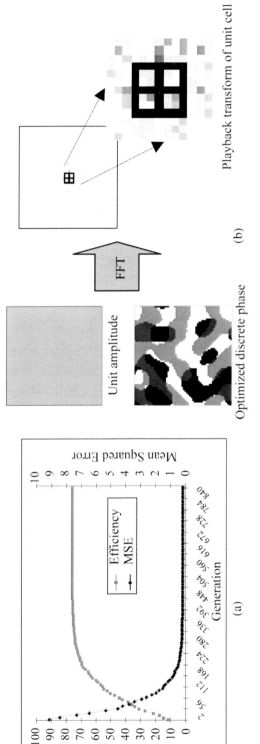

Figure 5.20 (a) The genetic algorithm reaches a minimum in about 400 generations. (b) The output of this design is similar to that of simulated annealing. The MSE is 0.11.

different, the nature of the design is similar. The design performance is also quite similar. The efficiency is 76%, which is nearly the same, and the MSE is 0.11. This is slightly better than the simulated annealing result, but it is too small a difference to state that one algorithm is superior to the other. Also, the total number of computations to reach our converged solutions was a little less than 10^{10}, so the computational price of this design was similar to that of simulated annealing.

We can attempt to relate genetic algorithms to our analogy of the blind prospector. To do this, we have to hire a team of blind prospectors scattered throughout the hills. Each can shout to the others about how far down the valley he is. Each prospector then leaps in the direction of the fellow he thinks is in the best position (see Fig. 5.21). The fellow in the truly best position may get worse, but several of his counterparts will get better by leaping toward him. None of their time is spent mapping the mountains since this would take a tremendous amount of time. Instead, our team of prospectors is equipped with three attributes that allow them to find good valleys. (1) They interact during each iteration. (2) They move stochastically, but with a goal. (3) They act as a team.

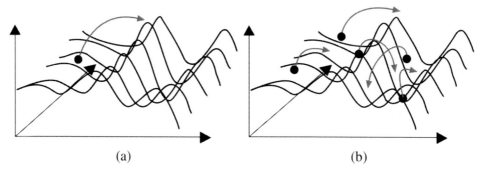

(a)   (b)

Figure 5.21 Approaches to optimization. (a) Simulated annealing permits the estimate to move "uphill" during the initial iterations. The algorithm thus finds better locations in the solution space. (b) Genetic algorithms use a population of estimates scattered throughout the solution space.

Genetic algorithms provide a robust technique to use in searching for solutions to systems with many variables. Like simulated annealing, the approach is very scalable. It will consistently converge to a minimum, but it is difficult to ensure that a genetic algorithm will arrive at the truly best solution or global minimum. Genetic algorithms are easy to understand because they mimic selective breeding, a familiar subject to many. We intuitively know that larger populations will bring greater diversity and better sample the solution space. If the ranking function [Eq. (5.21)] is nearly flat, poor attributes will stay in the population longer. If the ranking function is steep, the population swiftly becomes inbred. A lot of mutation can slow the convergence, while no mutation will lead to premature stagnation through inbreeding. The algorithm is easy to manipulate because of these intuitions.

Stochastic algorithms such as simulated annealing and genetic algorithms provide the best way to arrive at the best grating design. This performance comes at the computational cost of more calculation time. Both algorithms are scalable; indeed, they are designed to optimize systems of many variables. For many grating design problems, stochastic optimization algorithms are the method of choice.

Although this discussion of diffractive optics optimization has been confined to the design of diffraction gratings, these optimization techniques can be applied to nonperiodic structures also. The box describes one such example.

Example: A bidirectional algorithm for beam-shaping optimization

The challenge of reshaping a Gaussian laser beam to a uniform square is an excellent application of diffractive optics. The fields can be modeled mathematically as arrays of complex values, and a Fresnel transform can be calculated to simulate beam propagation. The Fresnel transform can be inverted. Although the number of variables is very large, the system has only one general solution, so the merit function in the solution space is smoothly varying.

The beam-shaping problem was framed as a bidirectional algorithm, using forward and inverse Fresnel transforms. The system had over 250,000 variables, and converged to a solution in less than an hour.

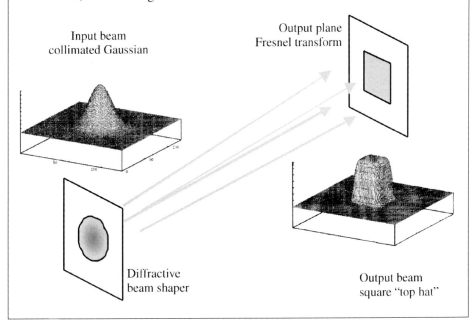

5.6 Conclusion

This chapter has discussed the characteristics of diffractive optics that perform $1 \times N$ mapping, as well as design techniques for creating them. The techniques are similar for both one-dimensional and two-dimensional gratings. The two underlying facts that simplify grating design are these: (1) The grating period alone controls the angular separation and orientation of the diffracted orders. (2) The structure within the unit cell of the grating alone controls the relative amplitude and phase of the diffracted orders. This separates the design procedure into two manageable tasks: First determine the grating period(s) from the spot separation in the desired output pattern and then design the unit cell to direct the light into the desired orders. While the calculation of the period is easily done analytically from the grating equation, the calculation of the unit cell structure can be significantly more involved. We have discussed a number of ways of encoding the unit cell, including transition locations and pixelated phase values. Although all of our design examples used a pixelated phase encoding, the algorithms can be extended to phase-transition encoding as well. The unit cell calculation can be as simple as direct inversion or as sophisticated as optimization of a genetic algorithm. Because the design technique employed is application dependent, the choice of algorithm should be based on the criterion of providing an acceptable design using the simplest technique.

References

1. H. Dammann and K. Gertler, "High-efficiency in-line multiple imaging by means of multiple phase holograms," *Opt. Commun.* **3**, pp. 312–315 (1971).

2. H. Dammann and K. Gertler, "Coherent optical generation and inspection of two-dimensional periodic structures," *Opt. Acta* **24**, pp. 505–515 (1977).

3. J. Jahns, M.M. Downs, M.E. Prise, N. Streibl, and S.J. Walker, "Dammann gratings for laser beam shaping," *Opt. Eng.* **28**, pp. 1267–1275 (1989).

4. A. Vasara et al., "Binary surface-relief gratings for array illumination in digital optics," *Appl. Opt.* **31**, pp. 3320–3336 (1992).

5. J.-N. Gillet and Y. Sheng, "Irregular spot array generator with trapezoidal apertures of varying heights," *Opt. Commun.* **166**, pp. 1–7 (1999).

6. J.-N. Gillet and Y. Sheng, "Iterative simulated quenching for designing irregular spot array generators," *Appl. Opt.* **39**, pp. 3456–3465 (2000).

7. H. Farhosh et al., "Comparison of binary encoding schemes for electron-beam fabrication of computer-generated holograms," *Appl. Opt.* **26**, pp. 4361–4372 (1987).

8. U. Krackhardt, J.N. Mait, and N. Streibl, "Upper bound on diffraction efficiency of phase-only fanout elements," *Appl. Opt.* **31**, pp. 27–37 (1992).

9. R.W. Gerchberg and W.O. Saxton, "Phase determination from image and diffraction plane pictures in the electron microscope," *Optik* **34**, pp. 275–284 (1971).

10. J.N. Mait, "Understanding diffractive optical design in the scalar domain," *J. Opt. Soc. Am.* **A 12**, pp. 2145–2158 (1995).

11. R.W. Gerchberg and W.O. Saxton, "A practical algorithm for the determination of phase from image and diffraction plane pictures," *Optik* **35**, pp. 237–246 (1972).

12. J.R. Feinup, "Iterative method applied to image reconstruction and to computer-generated holograms," *Opt. Eng.* **19**, pp. 297–305 (1980).

13. L. Ingber, "Simulated annealing: practice vs. theory," *J. Math. Comp. Modeling* **18**, pp. 29–57 (1993).

14. K.H. Hoffmann and P. Salomon, "The optimal simulated annealing schedule for a simple model," *J. Phys. A* **23**, pp. 3511–3523 (1990).

15. M.R. Feldman and C.C. Guest, "High efficiency hologram encoding for generation of spot arrays," *Opt. Lett.* **14**, pp. 479–481 (1989).

16. L. Davis, *Handbook of Genetic Algorithms*, Van Nostrand Reinhold, New York (1991).

17. Z. Michaelewicz, *Genetic Algorithms + Data Structures = Evolution Programs*, Springer-Verlag, Berlin (1992).

18. E.G. Johnson et al., "Advantages of genetic algorithm optimization methods in diffractive optic design," *Diffractive and Miniature Optics*, S.H. Lee, Ed., SPIE Press, CR49, pp. 54–74 (1993).

Chapter 6

Making a Diffractive Optical Element

In previous chapters we considered the fundamental theories of diffraction and a number of design approaches for diffractive optical elements. However, before a diffractive optical element can be built, a number of additional questions must be addressed. For example:

- How is the design in a computer "transformed" into a physical component?
- What material should the DOE be made from?
- How does the choice of material affect the fabrication process?
- Are there physical limitations on what can be made?
- How does the quality of the substrate affect the performance of the components?
- Are special pieces of equipment or special environments required for the fabrication process?
- How do you know what you have made?

These topics will be explored in detail in the next several chapters. In general, we will explore the manufacture and test of spatially varying phase profiles in or on an optical substrate. Although this phase profile can be achieved through a number of means, the most common realization of these profiles will be considered: Diffractive optical elements created as spatially varying surface relief profiles in or on an optical substrate.

6.1 The Profile

The surface relief profiles of diffractive elements can take on a wide range of shapes that depend on the optical function of the component. Consider the simple binary phase structure that was analyzed in Chapter 5 and is shown in Fig. 6.1. When the depth d is equal to a half-wave optical path difference (discussed later) and the duty cycle, $b/\Lambda = 1/2$, the grating is a square-wave type. This grating will split a normally incident light beam, with approximately 81% of the transmitted

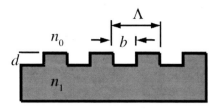

Figure 6.1 Geometry of a simple binary phase diffractive element.

light split equally between the positive first and negative first diffraction orders, as discussed in Sec. 2.4.1. As more elaborate and efficient diffractive structures are required, the complexity of the surface structure increases rapidly. Depending on the desired optical functionality, the required surface patterns might have two-phase steps, multiple-phase steps, or even be essentially continuous-relief surface profiles. Lateral patterns can be one or two dimensional, with varying degrees of symmetry or asymmetry. Submicron precision may be required for feature widths, and depth tolerances on the order of 10 nm or less may be needed. Simply stated, the fundamental problem in DOE manufacturing is: How do you fabricate these microscopic surface relief structures with the correct geometries?

Many different methods exist for fabrication of diffractive microstructures.[1] Most of these techniques can be grouped into three main categories: lithographic techniques, direct machining, and replication. Lithographic techniques (originally developed in the microelectronics industry) use light-sensitive polymers in conjunction with controlled etching or deposition methods. In direct machining, the surface relief structure is generated through direct removal of the optical material in a controlled manner without any intermediate processes. Replication techniques generate copies of surface relief structures in polymers or other materials from a "master" element. The choice of fabrication method is generally driven by two main factors: function and cost. Each method has various advantages and disadvantages. In many cases, the required performance of the DOE will inherently limit the fabrication methods that can be used. Photolithographic processing of diffractive optics using binary masks will be employed as the primary model for the discussion on fabrication in this text. A survey of other fabrication methods is presented in Chapter 8.

6.2 Photolithography: A Method for DOE Fabrication

In general, the most common processes for fabrication of diffractive optical elements involve photolithographic methods that borrow heavily from the microelectronics industry. They are based on the same processes used to fabricate integrated circuits.[2–5] Because the size of the features and the need for flexibility of pattern generation are similar to those in manufacturing semiconductor devices, lithographic methods are optimal for many types of DOE fabrication. These methods are based on the use of photoresist, a light-sensitive conformal polymer. A simpli-

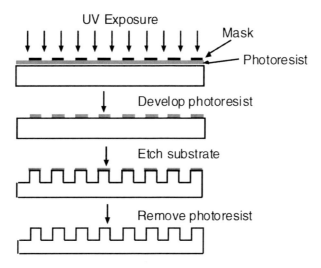

Figure 6.2 Photolithographic processing and etching.

fied version of the photolithographic processing sequence followed by an etching step is shown in Fig. 6.2. A more elaborate one will be presented in Chapter 7.

The next few chapters will focus on various aspects of lithographic fabrication and testing. A specific approach that generates up to 2^N phase levels from N binary transmission masks (commonly referred to as the *binary optics* method) will be used as an example throughout these chapters. We will examine issues related to mask generation, substrate testing, facilities, photoresist processing, etching into substrates, and DOE testing in detail.

6.3 From Equation to Component

Once the appropriate optical phase function is determined from the design process (discussed in earlier chapters), it is necessary to "translate" this phase function into a geometric surface relief pattern. In the scalar "thin-phase" approximation used for most diffractive optics, the surface profile can be split into two independent parts:

1. The lateral transition points in the pattern, which correspond to changes in the phase profile by 2π or some appropriate fraction of 2π.
2. The depth of the diffractive phase structure, which is determined by the wavelength of the incident light and the refractive indices of the substrate and surrounding material.

6.3.1 Converting function to form

As noted earlier, the geometry of the surface structure determines the function of the DOE. In some cases this can be represented by a relatively simple equation, while more complex DOEs must be designed numerically and iteratively on a computer, as discussed in previous chapters. As a simple example, consider the phase grating shown in Fig. 6.1. The period Λ of the grating determines the diffraction angles of the resulting diffraction orders, as given by the grating equation:

$$m\lambda = \Lambda(\sin\theta_m - \sin\theta_i), \tag{6.1}$$

where λ is the wavelength of the incident light, θ_i is the angle of the incident light wave, and θ_m is the diffraction angle of the mth diffraction order. Both θ_i and θ_m are measured from the normal to the grating plane, not the grating facets. Similarly, the relative phase difference imparted to the incident light wave is directly proportional to the depth of the local surface relief structure. The relationship between phase and depth is given by

$$d(x, y) = \frac{\phi(x, y)}{2\pi} \frac{\lambda}{(n_1 - n_0)}, \tag{6.2a}$$

for a transmissive DOE. For a reflective DOE, the depth is

$$d(x, y) = \frac{\phi(x, y)}{2\pi} \frac{\lambda}{2n_0}, \tag{6.2b}$$

where $\phi(x, y)$ is the phase in radians, and n_1 and n_0 are the indices of refraction of the substrate material and the surrounding medium at the operating wavelength, respectively. These simple equations "translate" the requirements for many diffractive optical elements from an optical function into a tangible design. We now consider a simple example.

6.3.2 Example: 1 × 2 beamsplitter

Consider the example of a transmissive binary phase grating designed to operate as a 1 × 2 splitter with an He:Ne laser ($\lambda = 0.6328$ μm) with an angle of 18.2 deg between the +1 and −1 diffraction orders. Using the grating equation with normally incident light, the period Λ of the grating is determined to be 4.0 μm. Because the duty cycle of the grating is 50%, the width of the feature b in such a grating is 2.0 μm. If the incident medium is air, and the substrate material is fused silica ($n = 1.4572$ at 0.6328 μm), then the depth of the structure is [Eq. (6.2)]

$$d = \frac{\pi}{2\pi} \frac{0.6328\mu m}{(1.4572 - 1)} = 0.6920\mu m. \tag{6.3}$$

A scanning electron microscope (SEM) image of this grating is shown in Fig. 6.3. Equations (6.1) and (6.2) also apply to more complex beam fanouts. The period of the grating structure determines the angular separation between the diffraction orders, while the structure *within* a single grating period determines how the power is distributed between the orders. Even in more complex cases, the depth of the grating is determined only by the incident wavelength and the indices of refraction.

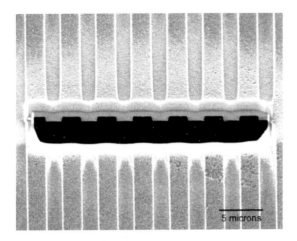

Figure 6.3 Cross-section of the 1 × 2 beamsplitter described in this example. (Courtesy of Digital Optics Corp., Charlotte, NC.)

Regardless of the fabrication method used, it is necessary to translate the data representing the physical structure of the DOE into a format that can be used to operate the fabrication machinery or process. As an example, we now describe how the masks are made for the photolithographic fabrication process shown in Fig. 6.2.

6.3.3 Mask generation

A number of different methods can be used to generate masks for the fabrication of diffractive optical elements. Most lithographic masks are binary transmission masks. That is, they consist of alternating clear and opaque areas. These masks are usually generated by forming a pattern in a light-sensitive photoresist on top of a thin chrome layer on the glass mask blank. Once the photoresist is exposed and developed, the chrome is etched away in the areas where the photoresist has been removed. The chrome remains where it was protected by the photoresist. After the remaining photoresist is removed, the chromed areas on the mask serve as the opaque regions that block the incident light, as shown in Fig. 6.2.

The mask pattern is exposed using optical pattern generators or electron-beam ("e-beam") machines. Those masks created by optical pattern generators are exposed by a flash lamp with a controllable aperture. More typically, masks are exposed by a raster-scanned electron beam. These electron beam writers are quite expensive; a new machine can easily cost more than $5 million. As a result, it is very rare for a manufacturing company or research lab to have its own e-beam writer. However, many "mask foundries" will generate lithographic masks from data files supplied by customers. Although these mask foundries are primarily supported by the microelectronics industry, masks for fabrication of diffractive optical elements can also be generated, provided the mask data representing the optical function are encoded in the proper format. A variety of file formats can be used to specify lithographic masks. These include the moving electron beam exposure system

(MEBES), graphic design system (GDSII), Caltech intermediate format (CIF), and data exchange format (DXF). Most mask foundries use e-beam writers that require the MEBES format. Significant software development is required for the generation of files in this format; most DOE fabricators have proprietary algorithms for creating mask data files that are optimum for diffractive patterns.

The MEBES format is a concise raster format for mask data files and is commonly used for making masks for integrated circuit manufacturing. The smallest divisible unit of the MEBES format is the address unit, which is similar to a pixel on a computer monitor. Stripes and blocks are also units of spatial measurement used in MEBES writing (Fig. 6.4). Each stripe is defined as a 32768×1024 raster of address units. These units represent the area in which the electron beam can be electronically scanned without physical movement of the mask blank on the translation stages (modern MEBES machines can write a 4-stripe $\times$ 1-block area without requiring stage motion). Finite translational accuracy of the movement stages can result in butting errors, which are very small discontinuities (on the order of nanometers) between the stripes. The mask patterns are "fractured" into rectangles and trapezoidal shapes, also referred to as primitives, within a given stripe. As a consequence, e-beam systems are best suited for mask patterns requiring rectilinear geometries. This puts special requirements on masks for diffractive optical elements needing curved patterns, such as diffractive lenses. After the data conversion process is completed, the mask pattern is written one address unit at a time by a raster exposure within each stripe.

The accuracy of a curve fit by these primitive shapes is directly related to the wavefront error introduced by the required shape approximation. The local wavefront error in a mask pattern is $\approx \varepsilon \sin \theta / \lambda$, where ε is the fit error and θ is the local

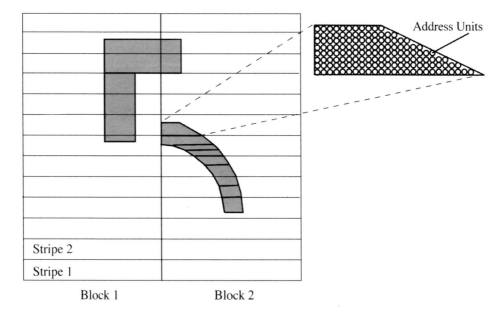

Figure 6.4 Concepts for mask patterns in MEBES format.

diffraction angle. Increasing the number of trapezoidal shapes and using smaller address units improves the fit to the ideal mask pattern, particularly when curved patterns are required. However, as one might expect, the higher degree of accuracy obtained through both of these methods does not come for free. The increased costs associated with more trapezoids resulting from data flow issues can be significant. Similarly, patterns with smaller address units require more time to write the pattern and thus cost more. A relatively simple pattern written with 0.5-μm address units might cost on the order of $1000, while a very complicated mask written with 0.1-μm address units can cost $20,000 or more. For this reason, the degree of wavefront accuracy required should be examined very closely during the mask layout process. As an example, a four-inch patterned area written at a 0.2-μm address unit constitutes approximately 2.6×10^{11} data points. This amount of data is equivalent to about 38,000 letter-sized pages printed at 300 dots per inch!!! Modern MEBES machines write approximately 10^7 to 10^8 address units per second, depending on the complexity of the pattern.

6.4 Interplay between Fabrication and Optical Design

Fundamental issues and methods of diffractive optical design have been discussed in detail in earlier chapters. However, many of the choices made in realizing the design can have a significant impact on the fabrication process. A clear understanding of the limitations of the fabrication process and substrate material is crucial to the successful fabrication of a DOE. We now consider several of these issues.

6.4.1 Optical functionality

In many cases, the function of the DOE will limit the fabrication methods that can be used to make the piece. For example, direct machining by ruling engines or diamond-turning machines is limited to producing structures to either straight-line or circularly symmetrical gratings, respectively. It would be extremely difficult (if not impossible) to fabricate a generalized two-dimensional spot array generator or diffractive beam shaper using either of these methods. Conventional holographic methods can produce complex optical functions, but realization of arbitrary functions is difficult because of the inability to create the required object and reference beams. Methods of computer-generated holography provide the ability to create essentially arbitrary surface geometries, and these can be fabricated using lithographic methods. Although lithographic techniques provide the greatest flexibility in choice of optical structures that can be fabricated, even they have their limitations.

6.4.2 Fabrication constraints

Any limitations that are due to the chosen fabrication method must be understood before the design of a DOE is completed. The most common fabrication constraint

PostScript Methods for Mask Generation

Inexpensive lithographic masks of moderate quality can also be made using methods of desktop publishing. These masks are made using a different pattern generator, a high-resolution, laser-based imagesetter. These devices are used by graphic-arts houses to generate printing plates for publishing. Although graphics houses are not mask foundries and their staffs have no idea how to make a diffractive optic, they *do* accept standard graphics files and can print the patterns when the graphics are described in PostScript, a standard page composition language. Using desktop publishing programs and producing a pattern description in the PostScript language can provide a low cost method for fabrication of diffractive masks with moderate resolution. Letter-sized transparencies can be made for about $25 in less than one day.

PostScript is a device-independent programming language that describes graphics and text with a large number of primitive commands. The same commands to a laser printer and an imagesetter will produce the same image, but with different resolutions.

Although the PostScript language can be used directly for shape generation, other software packages are normally used to generate the PostScript code. For example, both *Mathematica*, a mathematical calculation program, and CODE V, an optical design program, can generate PostScript files for printing. However, neither package will be available in the typical graphic arts house. So, the patterns from these applications are imported into a commonly used PostScript-based graphic illustration programs such as Macromedia Freehand or Adobe Illustrator, which are widely used in the graphic-arts industry. Many of the characteristics and features of these programs make them ideal for mask generation. In some cases it is possible to draw very elaborate unit cells within the graphic illustration program and use a repeat generation feature of the program to draw the grating without having to align any of the unit cells.

Although a high-resolution mask from an imagesetter might contain features as small as 30 μm, this is still too coarse for most applications of diffractive optics. In addition, the film on which the pattern resides is not rugged enough to withstand the handling required to transfer the pattern to photoresist. For these reasons, masks with smaller features can be produced by photoreducing the pattern on the transparency film onto a glass film plate. Diffractive optics with minimum feature sizes of approximately 7 μm have been fabricated using this approach. More details can be found in several references.[5-9]

is the minimum feature size (mfs). It is surprisingly easy to design a diffractive optical element that requires features that cannot be fabricated. The minimum feature size depends on the fabrication process used to make the structure. If the limitations of the chosen fabrication method are known in advance, it may be possible to maximize the performance of the optical element subject to the fabrication constraint. In most cases, the minimum desired feature size can be set as a constraint in an iterative design optimization process, resulting in an optimal solution.

Another key fabrication constraint to be considered during the design process is the aspect ratio of the microstructures. The aspect ratio is the height of a given microstructure divided by its width, d/b, as shown in Fig. 6.1. As a rule of thumb, the higher the aspect ratio of a given structure (i.e., the skinnier the feature), the more difficult it will be to manufacture accurately. For lithographic processes, fabrication of deep structures requires thicker photoresist layers to protect the substrate during etching. However, thicker photoresist layers are more susceptible to diffractive spreading of the incident light during exposure, reducing the fidelity of the patterned microstructure. A common practice in establishing manufacturable designs is to keep the aspect ratio to less than one. However, structures with aspect ratios of 5 to 10 or more can be made with special effort by using special etch chemistries or etch stop materials other than photoresist (such as metals or thin dielectric layers) with appropriately chosen etch chemistries. This approach is discussed in more detail in Sec. 7.1.4.

Example: 1 × 5 Dammann Grating

One solution for a binary phase 1 × 5 Dammann grating contains a minimum feature that is approximately 7% of the width of the grating period. Consider the case where an angle of 7.5 deg is desired between each of the diffraction orders when the grating is illuminated with light at 1.310 μm at normal incidence $\theta_i = 0$. From the grating equation [Eq. (6.1)], the necessary grating period Λ is 10 μm. Consequently, a 0.7-μm feature is required in each of the grating periods. The depth of such a grating in fused silica is 1.48 μm [see Eq. (6.2)]. This feature dimension is within the range of contact lithography and projection lithography if electron beam-generated masks are used, but it is beyond the capability of PostScript-based masking methods. The aspect ratio for the smallest feature in the grating (more than 2:1) is challenging but achievable.

6.4.3 Effects of thin-film coatings

In many cases, it is desirable to overcoat a diffractive optical structure with reflective metals or dielectric thin films to enhance or reduce reflections from the optical surface. Diffractive optical elements exhibit Fresnel reflection losses in much the same manner as refractive surfaces; a glass diffractive element in air will reflect approximately 4% of the incident light per surface, much as a silicon diffractive element will reflect approximately 30% of the incident light per surface. Thus, in order to maximize the overall efficiency of a transmissive diffractive optic, it is desirable to minimize the amount of reflected light. While the use of antireflection (AR) coatings with refractive surfaces is relatively straightforward, their use with diffractive optical elements requires a bit more care and forethought. These films can be considered "thin" for standard macroscopic optical components, but the thickness of the "thin" film can be a significant fraction of the dimensions of a diffractive structure. Furthermore, the material is deposited on both the "top" of the diffractive structure as well as on the side walls. As a consequence, the addition of a dielectric layer or stack can significantly change the performance of the diffractive structure, particularly if the feature sizes in the diffractive structure are small. A scanning electron microscope (SEM) cross-section of a simple diffractive structure with an AR coating is shown in Fig. 6.5.

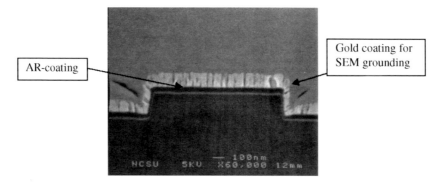

Figure 6.5 Cross-section of diffractive structure with AR coating on top. (Courtesy of Digital Optics Corp., Charlotte, NC.)

Because the coating is deposited on both the top and the sides of the diffractive structure, competing effects may be observed. The use of the AR coating can significantly reduce the amount of light reflected (thus enhancing the transmission), but the energy directed into a desired diffraction order(s) may not improve at all; in fact, in some cases the performance can be made worse by the addition of the AR coating. Side wall deposition changes the grating profile (thus changing the light distribution in the diffracted orders) without increasing the transmission of the component! Thus, if thin-film coatings are to be used, it is advisable to perform a detailed analysis (if feasible) to determine if the addition of the AR coating will help or hurt the performance of the diffractive component. Still, very high-performance diffractive optical elements with AR coatings have been demonstrated.[11]

There are multiple deposition methods [sputtering, evaporation, ion-assisted deposition (IAD), etc.] with many different process parameters that affect the coating characteristics.[12] Ultimately, the effect of the coating on device performance is dependent on the type of coating, the deposition method used, the deposition parameters, and the physical geometry of the diffractive structure itself. As a general statement, anisotropic deposition processes (in which the deposited layers are preferentially deposited on the planar surfaces of the diffractive as opposed to the side wall of the structure) are preferred.

6.4.4 Materials

The choice of substrate material during the design process is typically driven by the spectral transmission properties as well as the refractive index of the material. For reflective elements, the reflectivity of the material or coating is the key concern. The coefficient of thermal expansion of the material can also play a major role in choice of material. Several common materials used for the manufacture of micro-optics are listed in Table 6.1, along with the approximate spectral transmission bands and refractive indices of these materials. More detailed information on the optical properties of materials can be found in several references.[13,14]

If there is a choice, materials with high refractive indices can be desirable as DOE substrates since the depth of the structure needed to achieve a particular phase difference at a chosen wavelength is smaller than the depth needed for the equivalent structure in a material with a low refractive index. Because the structure is shallower, its aspect ratio is smaller, and shadowing effects that can reduce the overall efficiency of the diffractive element are minimized, as discussed in Sec. 2.5. However, a large refractive index also places stricter tolerances on the etch depth of the diffractive element. From Eq. (6.2), a 1% phase error in fused silica at 1.55 μm ($n = 1.444$) for a binary phase element corresponds to a depth error of about 17.5 nm, while a 1% error in silicon for the same case corresponds to a depth error of about 3 nm! Materials with large refractive indices also have much higher reflective losses; the total light transmitted through a slab of fused silica at normal incidence is approximately 92%, while a silicon wafer transmits only about 49% of the incident light. All these factors should be taken into consideration when choosing the material for a given diffractive element.

In addition, the choice of substrate material can also have a profound impact on manufacturability. In particular, this choice plays a significant role in the method used to etch the structures into the substrate. Different materials may require significantly different etching methods. Methods for etching substrate materials for diffractive optics are discussed in more detail in Chapter 7.

6.5 Facilities and Substrates

Before fabrication of a diffractive optical element can begin, the substrate must be properly prepared. This preparation and the masking and etching steps that follow must be done in a highly controlled environment. Although it is possible to fabricate diffractive optical elements in laboratory environments, volume production of micro-optics requires equipment and facilities comparable to those found in micro-electronics manufacturing facilities. We describe here the supporting infrastructure

Table 6.1 Optical materials for diffractive optical elements.

Material	Approximate λ range	Refractive Index	Comments
BK7	350 nm to 2 μm	~ 1.5	Impurities in the glass can make it very difficult to fabricate precision microstructures
Fused silica	150 nm to 3 μm	~ 1.5	Material of choice for DOEs at visible wavelengths, near-UV, and near-IR
Si	1 to 7 μm	~ 3.5	Good for infrared optics; high refractive index can be both good (shadowing) and bad (tight tolerances on depth control)
ZnSe	0.6 to 18 μm	~ 2.4	Commonly used for night vision systems and with CO_2 lasers
ZnS	4 to 13 μm	~ 2.2	Commonly used for night vision systems and with CO_2 lasers
Ge	2 to 20 μm	~ 4.0	Typically used in mid-IR
GaAs	1.5 to 18 μm	~ 3.3	Typically used in mid-IR. Useful for integration with active functions
CdTe	1 to 30 μm	~ 2.7	Used with infrared focal plane arrays for material compatibility
InP	1 to 10 μm	~ 3.3	Can be used for both active and passive optical functions, but difficult to process
CaF_2	150 nm to 7 μm	~ 1.4	Useful from the UV to the IR, but challenging to fabricate DOEs in this material
Plastics	400 to 1600 nm	~ 1.4 to 1.6	Many different types. Inexpensive, but generally have reduced optical and environmental performance relative to other materials

that must be in place first and then the steps needed to prepare the substrate for processing.

6.5.1 Clean rooms and DOE fabrication

Cleanliness is a necessity for fabrication of precision DOEs. Any type of particulate or chemical contamination on the substrate or lithographic mask has a negative impact on the fabrication (and ultimately on the performance) of a diffractive optical element. For this reason, it is advisable to fabricate DOEs in a clean room environment. A clean room is a controlled space with filtered air and environmental controls. The cleanliness factor of a clean room is normally given in terms of a class number (class 10, class 1000, etc.), which refers to the average number of particulates greater than 0.5-μm diameter contained in one cubic foot of air. The requirements for DOE fabrication need not be as stringent as in microelectronics manufacturing, which can require class 1 clean room facilities. Class 1000 environments are sufficient for fabrication of most DOEs. However, accurate, repeatable fabrication of DOEs (particularly with small features) may require class 100 environments or better.

Clean room air may be cleaned in a number of ways. In most manufacturing facilities, the entire room contains clean, filtered air. Air enters the room through high-quality high-efficiency particulate air (HEPA) filters in the ceiling and is removed through output vents on the walls. The entire room is also maintained at a positive pressure, meaning that the overall atmospheric pressure in the room is higher than that in surrounding areas. Since air naturally moves from areas of high pressure to areas of low pressure, air (and thus dust particles and other contaminants) cannot enter the room from outside areas except through the filters. In some cases the air in the room is only "clean" in certain areas; all key fabrication processes are performed under laminar flow hoods that provide clean, filtered air in a controlled area.

Temperature and humidity must be controlled for precision fabrication of microstructures because some lithographic processes are dependent on both conditions. For example, photoresist adhesion and the development rate of photoresist are affected by moisture. In addition, temperature variation causes changes in the rates of chemical processes and can also introduce significant problems from thermal expansion or contraction of lithographic masks or equipment components. For precision lithographic work, the operating temperature of the room should be the same as the temperature at which the lithographic mask was generated at the mask foundry. In manufacturing clean rooms the temperature is controlled to about $68 \pm 1°$F and the relative humidity is controlled to $45 \pm 5\%$. Some pieces of manufacturing equipment may have additional self-contained environments for even tighter temperature and humidity control. In a field where submicron pattern fidelity and registration is required, environmental controls are a must!

While specific layouts of clean rooms will vary according to equipment, processing volume, and many other factors, there are several areas that are typically found in clean rooms intended for fabrication of diffractive optical elements.

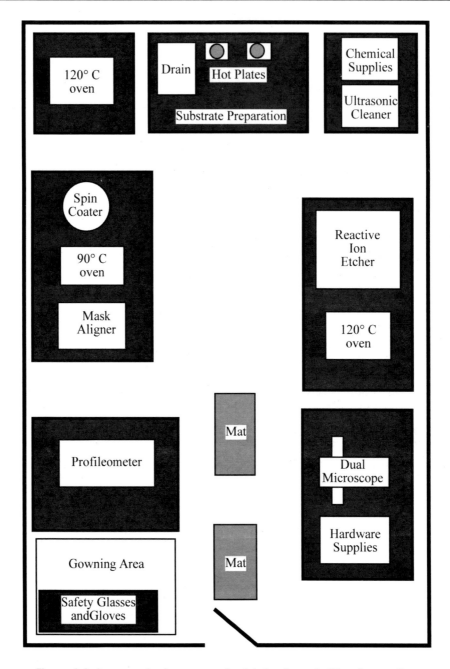

Figure 6.6 A research clean room for fabrication of diffractive optics.

These areas are described in the following paragraphs. A sample layout of a research clean room for diffractive optics fabrication is shown in Fig. 6.6.

Gowning Area: This is a location for putting on gloves and clean room garments (affectionately referred to as "bunny suits" in many facilities) to prevent contamination of the room by clothing or skin oils, as well as protective goggles and other safety apparatus. The gowning area is typically segregated from the main body of the clean room (or is in another room altogether) and is separated further by a series of adhesive floor mats that remove dirt and dust from shoes and shoe coverings.

Substrate Preparation Area: This area is typically located under a fume hood for removal of harmful chemical vapors. It is the location where chemical solvents may be safely used for cleaning the substrate before further processing. Wet chemical development of photoresist after exposure may also be performed in this area.

Lithography Area: This area will contain the tools necessary to fabricate microstructures in or on the optical substrate. Typical tools will include a spin coater for coating the substrates with photoresist, curing ovens or hot plates for photoresist baking, an exposure and/or alignment tool, and an etching tool such as a reactive ion etcher (RIE) for transferring the patterned structure into the substrate.

Inspection Area: This area is used to track and verify the fidelity of the different stages of the fabrication process. It will typically contain an optical microscope and a mechanical surface profilometer. In some cases, other equipment may also be used, including thin-film measurement devices, atomic force microscopes (AFMs), scanning electron microscopes (SEMs), and other pieces of equipment for process metrology.

To maintain a high level of cleanliness in a clean room, the transfer of particle and chemical contaminants must be eliminated wherever they are found. It is enlightening to consider major sources of contamination of a work piece:

1. Transport boxes
2. Process equipment
3. Tweezers and other handling devices
4. Nearby surfaces during handling
5. Operators (the biggest source of contamination!)

Generally, the clean-room user is the largest source of contamination in the entire room! Dust, hair, skin oils, and even moisture from breathing introduce contamination into the room. With this in mind, it is useful to adopt a "jelly donut" mentality when entering a clean room. When you eat a powdered jelly donut, you must be careful how you handle it, since the powdered sugar can spread to everything: your hands and face, your clothes, other people, and anything else you come in contact with or that you pass over. In a clean room, think of yourself as a giant powdered jelly donut. Although part of you may be covered by a gown, hood, mask, and gloves, there are still uncovered places that can spread powdered sugar. If you are aware of your dual role in this process (as both fabricator and contaminator), you will operate differently than if you are just a fabricator. Attitude toward cleanliness ultimately determines the success of any clean room policy. All of the expensive equipment installed to provide a controlled atmosphere is worthless if each of the users does not understand and follow the policies.

6.5.2 Substrate testing and cleaning

In addition to ensuring the dimensional integrity of the masks that will be used during fabrication, it is necessary to ensure that the substrate itself does not distort a transmitted wavefront. The amount of aberration in the substrate will be carried forward, along with the desired phase differences arising from the fabricated profile, with potentially adverse effects on optical performance. These aberrations, called *transmitted wavefront errors*, can be introduced by thickness variations, material inhomogeneities that cause variations in the refractive index, and other optical errors. Besides bulk effects, the wafer must also be very flat for high-resolution lithographic patterning, particularly for those lithographic processes where the mask is placed in contact with the substrate being patterned ("contact printing"), and for projection lithography of extremely small features in which the imaging system has a small depth of focus. Typical substrate sizes for diffractive optics range from diameters of about 20 mm up to 150 mm or more, with thicknesses ranging from a few hundred microns to several millimeters. As a general rule, better flatness and less transmitted wavefront error are achieved with smaller, thicker substrates, although large, thin substrates of acceptable quality can be obtained from wafer polishing vendors at higher cost.

Phase-shifting interferometry is commonly used to measure wafer flatness as well as transmitted wavefront error. The interference pattern generated by two optically flat surfaces tilted at a small angle to one another will also result in a series of parallel fringes. Deviations from straight-line fringes are used to determine the optical errors in the piece being tested. For example, if a substrate piece is inserted between the optical flats, any transmitted wavefront errors caused by the substrate will distort the straight fringes, owing to the flats (see Fig. 6.7). A computer program analyses the fringe pattern and displays a topographic map of the surface in units of wavelengths. From these data, the maximum wavefront error from peak to valley across the wavefront contour can be computed. Another measure gives the root-mean-square (rms) wavefront error over the same contour. Typical substrates for fabrication of diffractive optics range from $\lambda/4$ to $\lambda/20$ rms variation. A simple test geometry and a fringe pattern for a "good" substrate are illustrated in Fig. 6.7.

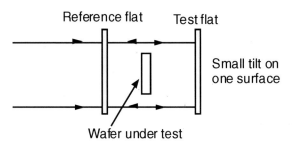
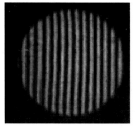

Figure 6.7 (Left) Measurement setup for transmitted wavefront error. (Right) fringe pattern of a "good" wafer.

Once the substrate quality and flatness have been ensured through interferometry, the substrate must be cleaned thoroughly for lithographic processing. The appropriate level of cleanliness is achieved by (1) subjecting the substrate to a rigorous series of cleaning procedures and (2) performing the lithographic steps in the clean room environment described in Sec. 6.5.1. This is because the quality of a DOE depends not only on preserving detail in the mask, but also on preventing extraneous details from contamination that could blemish or distort the desired pattern. A thorough cleaning of the substrate is strongly recommended before any fabrication procedure is started. Although it would be helpful to be able to specify what a "clean" surface consists of, unfortunately the specification has to be stated in the negative. That is, the surface should have no contamination from any oils, dust, or films. It has been found in medicine and in technology that it is easier and less costly to perform a lengthy set of cleaning procedures with high-grade materials than to subject a surface to contamination analysis methods, especially when the use of these procedures could introduce to additional sources of contamination.

To ensure photoresist adhesion, an extremely clean, dry substrate is needed. Dust particles and other chemical impurities are removed by rinsing the surface of the substrate with a series of solvents. Typical cleaning agents include acetone, methanol, and isopropanol. This series of organic solvents will remove most organics and oils from the substrate surface. Another common wafer cleaning sequence is referred to as the "RCA method" after the location where it was developed. The first rinse (RCA-1) removes organic contamination with a mixture of de-ionized water, hydrogen peroxide, and ammonium hydroxide. Metal contaminants are removed by the RCA-2 rinse, which consists of a mixture of de-ionized water, hydrogen peroxide, and hydrochloric acid.

The use of N-methyl pyrrolidinone (NMP) or an oxygen plasma to remove photoresist and organic contaminants from the surface of the substrate is very common. "Piranha" cleaning with mixtures of sulfuric acid, hydrogen peroxide, and peroxydisulfuric acid is also used. Afterward, the substrate must be dried thoroughly since water films, which adhere well to fused silica, will affect the development of the photoresist. So, after cleaning, the substrates are baked on a hot plate or in a dehydration oven to evaporate any moisture on the surface.

6.6 Fabrication of DOEs

We have discussed the generation of lithographic masks, substrate needs, and the facilities issues related to the fabrication of diffractive optical elements. With these things in place, we are now ready to examine the lithographic processes used for the fabrication of DOEs. The application of these processes to the fabrication of a multilevel diffractive optical element will be presented as an example of the process technologies in Chapter 7.

References

1. T.J. Suleski, "Diffractive optics fabrication," in *Encyclopedia of Optical Engineering*, R.G. Driggers, Ed. Marcel Dekker, New York (2003).
2. P. Rai-Choudhury, *Handbook of Microlithography, Micromachining, and Microfabrication, Vol. 1: Microlithography*. SPIE Press, Bellingham, WA (1997).
3. J.R. Sheats and B.W. Smith, *Microlithography Science and Technology*. Marcel Dekker, New York (1998).
4. H.J. Levinson, *Principles of Lithography*. SPIE Press, Bellingham, WA (2001).
5. S. Wolf and R.N. Tauber, *Silicon Processing for the VLSI Era*, Vol. 1: *Process Technology*, Lattice Press, Sunset Beach, CA (1986).
6. A. Nelson and L. Domash, "Low cost paths to binary optics," in *Conference on Binary Optics: An Opportunity for Technical Exchange*, H.J. Cole and W.C. Pittman, Eds. NASA **3227**, pp. 283–302 (1993).
7. D.C. O'Shea, J.W. Beletic, and M. Poutous, "Binary-mask generation for diffractive optical elements using microcomputers," *Appl. Opt.* **32**, pp. 2566–2572 (1993).
8. T.J. Suleski and D.C. O'Shea, "Fidelity of PostScript-generated masks for diffractive optics fabrication," *Appl. Opt.* **34**, pp. 627–635 (1995).
9. T.J. Suleski and D.C. O'Shea, "Gray-scale masks for diffractive-optics fabrication: I. Commercial slide imagers," *Appl. Opt.* **34**, pp. 7507–7517 (1995).
10. D.C. O'Shea and W.S. Rockward, "Gray-scale masks for diffractive-optics fabrication: II. Spatially filtered halftone screens," *Appl. Opt.* **34**, pp. 7518–7526 (1995).
11. E. Pawlowski and B. Kuhlow, "Antireflection-coated diffractive optical elements fabricated by thin-film deposition," *Opt. Eng.* **33**, pp. 3537–3546 (1994).
12. M.J. Madou, *Fundamentals of Microfabrication: The Science of Miniaturization*, 2nd ed. CRC Press LLC, Boca Raton, FL (2002).
13. *The Photonics Design and Applications Handbook*, Book 3, Laurin Publishing, Pittsfield, MA (2001).
14. E. D. Palik, Ed., *Handbook of Optical Constants in Solids*, Academic Press, San Diego, CA (1997).

Chapter 7

Photolithographic Fabrication of Diffractive Optical Elements

7.1 Photolithographic Processing

The photolithographic processes used for fabrication of diffractive optical elements adapted from the microelectronics industry were broadly described in the previous chapter. These methods are based on the use of photoresist. A more detailed photoresist processing sequence is shown in Fig. 7.1. In this chapter we examine the principles of photolithography and etching and explore their application to the fabrication of diffractive optical elements using binary lithographic masks. Additional

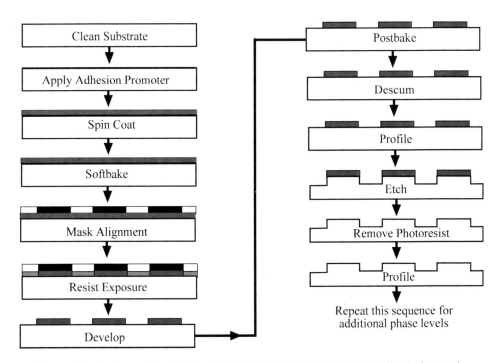

Figure 7.1 Photoresist processes for fabrication of diffractive optical elements.

133

information on the principles of photolithography can be found in a number of texts.[1-7]

7.1.1 Photoresist coatings

As indicated, photolithographic methods are based on the use of photoresist to create relief structures on substrate surfaces. This structure is used as a masking material to protect the underlying substrate during subsequent processing steps. In some cases, the photoresist structures themselves can be used as diffractive optical elements. Photoresists can be either positive, where the exposed resist dissolves upon development, or they can be negative, where the exposed resist polymerizes and remains after development. Positive photoresists are the most common type and are used in the processing examples given in this text.

There are many types of photoresists, each of which has been optimized for different applications. Most photoresists consist of three main chemical constituents: the photoactive compound (PAC), the solvent carrier, and the matrix material in which the PAC and solvent are contained. The PAC is the "light-sensitive" component of the photoresist and is typically designed to give optimal response at a specific exposure wavelength. The PAC in positive resists is typically diazide naphthaquinone, or DNQ. The solvent keeps the resist in a liquid state during storage and initial processing steps. The matrix material (usually a novolak resin) gives the photoresist its mechanical properties and serves as the protective layer after exposure and development. The reaction to incident light determines the quality of the three-dimensional photoresist pattern created from the two-dimensional lithographic mask. Different resists respond differently to different exposure wavelengths. This behavior can be quantified though three response parameters (A, B, C), referred to as the *Dill parameters* after a pioneer in the field of lithography modeling. The Dill parameters are defined as follows:

- *Absorption parameter A* ($1/\mu$m): This describes the bleachable absorption of the resist. A positive value means the resist bleaches (positive resist) and a negative value implies that the resist darkens. This is the absorption of the photoresist before exposure.
- *Absorption parameter B* ($1/\mu$m): This gives the absorption after the photoresist has been exposed and bleached. This is the nonbleachable part of the resist absorption. The thicker the layer of the photoresist and the larger parameter B is, the less light will reach the layers closer to the substrate surface.
- *Rate constant C* (cm^2/mJ): This parameter describes how fast the photoresist is exposed.

Graphs of the A and B Dill parameters for a positive photoresist are shown in Fig. 7.2.

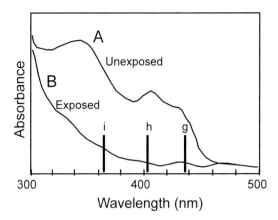

Figure 7.2 Sample graph showing absorbance properties for exposed and unexposed photoresist.

7.1.2 Spin coating photoresist

After a substrate is cleaned, the first step of the photolithographic process is coating the substrate with a thin (typically microns or less) layer of photoresist. Thin, uniform coatings of photoresist are generated by spin coating the wafer. In many cases, adhesion promoters such as hexamethydisalizane (HMDS) are applied to the clean wafer surface before spin coating to improve the adhesion of the photoresist layer to the substrate. During spin coating, liquid photoresist is dispensed on a substrate as it rotates at high rates (typically thousands of revolutions per minute). As the liquid resist spreads across the surface, the solvents in the resist begin to evaporate, causing the photoresist to slightly harden so that a skin begins to form on top (Fig. 7.3).

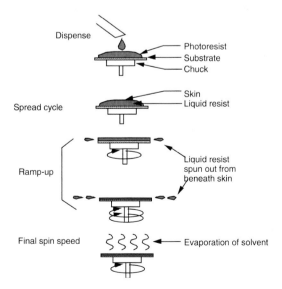

Figure 7.3 Spin coating photoresist onto substrates.

The final thickness of the photoresist layer is determined by a combination of the viscosity of the photoresist and the spin speeds used during the coating process. After spin coating, the wafer is baked in an oven or on a hot plate to drive out more of the solvent carrier. This step is sometimes referred to as the softbake and is a critical step in resist imaging. The resulting thicknesses obtained from a variety of Shipley photoresists as a function of final spin speed are shown in Fig. 7.4.

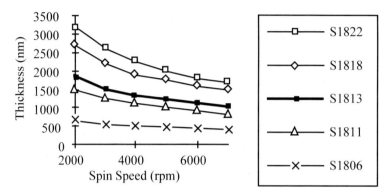

Figure 7.4 Photoresist thicknesses as a function of final spin speed for several similar photoresists.

D.J. Elliot noted in his basic text on photolitography that a good photoresist coating is critical for high-fidelity lithography.[*] If a particle gets into the photoresist, it will cause an uneven flow of resist around the obstruction, resulting in a "comet-tail" pattern. Other types of nonuniformities in the resist coating, referred to as striations, can be minimized by adding chemical "leveling agents" to improve coating uniformity.

7.1.3 Exposure and development

The next step in the lithographic fabrication of a diffractive optical element is the exposure of the photoresist. Patterns are formed in the photoresist layer using a spatially varying pattern of light energy created with a lithographic mask in conjunction with a uniform ultraviolet light source. After exposure, the substrate is subjected to a development step in which the areas of the (positive) photoresist layer that were exposed are washed away. For a positive resist, exposure to UV light initiates a chemical reaction that creates carboxylic acid in the exposed region and releases nitrogen gas. Developers for DNQ resists are aqueous, basic solutions such as NaOH, KOH, or tetramethyl ammonium hydroxide (TMAH). Applying the developer to the exposed piece removes the photoresist from the exposed area. The exposure and development processes are shown for a positive photoresist in Fig. 7.5.

---

[*]"The spin coating step, preceded by proper surface preparation, should be regarded as a foundation upon which the entire lithography and pattern transfer process is based."[2]

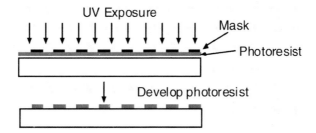

Figure 7.5 Photolithographic exposure and development for a positive photoresist.

The vast majority of photolithographic processes for integrated circuits and diffractive optics are performed using "binary" lithographic masks with clear and opaque regions, usually chrome on a glass mask. The mask can be placed in intimate contact with the photoresist layer for a high-resolution, 1:1 transfer of the image scale; this process is referred to as *contact printing*. A similar method that increases the lifetime of the lithographic mask at the cost of patterning fidelity by not bringing the mask and photoresist into intimate contact is referred to as *proximity printing*. It is also possible to project an image of the lithographic mask onto the photoresist layer using *projection lithography*. With the proper arrangement of high-quality lenses, 1:1 imaging can be achieved, or more typically, photoreductions of the mask geometries by factor of 5:1 or 10:1. Very small feature sizes can be achieved using both contact printing ($\sim$0.5 μm) and projection lithography ($<$0.3 μm).

Step-and-repeat projection lithography systems, or "steppers" for short, combine these lens arrangements with high-speed precision-movement stages and automated substrate handling. Modern steppers can expose more than a hundred wafers per hour with high precision and process repeatability. Historically, contact lithography has been the primary method used for fabrication of diffractive optical elements, but the use of stepper systems is becoming increasingly more common as both the technical specifications and the required production volumes for diffractive optics become more demanding.

A property of photoresist that is relevant to the fabrication of DOEs is the photoresist response. This relates the amount of photoresist remaining as a function of exposure energy for a fixed development time. The exposure energy (in millijoules per square centimeter) is obtained by multiplying the power rating of the exposing lamp (in milliwatts per square centimeter) by exposure time (in seconds). A sample response curve for a positive photoresist is shown in Fig. 7.6. An important point on the response curve is the *dose to clear* (E_0), the dosage above which all photoresist will be removed by the developing step. The other important point is the "knee" of the curve, that is, where the curve turns downward. The slope of the curve beyond the knee of the curve is a measure of the contrast of the photoresist. The ideal curve for producing lithographic patterns with sharply defined interfaces has a sharp drop at the knee, a characteristic of a "high-contrast" photoresist.

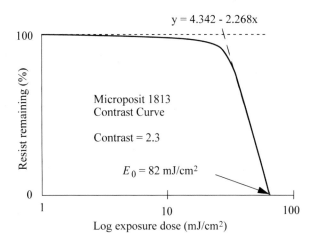

Figure 7.6 A sample photoresist response curve.

For binary lithographic masks, only two points on the response curve are used in a given exposure: the low-exposure energy area corresponding to the opaque areas on the mask and the high-exposure energy area above E_0 corresponding to the clear mask regions. Some methods of diffractive optic fabrication, such as direct writing and gray-scale lithography (discussed in Chapter 8), make use of the photoresist response curve between the knee of the curve and E_0 to create a diffractive structure with many levels using a single exposure and development step. These processes benefit from "low-contrast" (gradual slope) photoresists. Exposure and development parameters must be tightly controlled for these types of processes since small variations can significantly alter the resulting resist profile.

Substrates with high reflectivity, such as materials with a high refractive index, or substrates that have been coated with metal before exposure, can introduce the need for additional processing steps. During exposure of such substrates, a standing wave will be established in the resist layer because the reflected wave interferes with the incident wave, causing a corrugated side wall after development. This effect can be eliminated by performing a postexposure bake after exposure and before development to smooth the chemical gradients in the resist that correspond to the varying light density from the standing wave. Prolith simulations illustrating the effects of standing-wave interference in a photoresist layer and the smoothing effects of the postexposure bake are shown in Fig. 7.7. Another method to reduce or eliminate the standing-wave effect is to coat the substrate with an antireflection coating (ARC) before coating the substrate with photoresist to reduce the reflectivity off of the substrate. The ARC can be deposited as a dielectric thin film, or in many cases, can be created by spin coating a thin layer of organic material of the appropriate thickness and refractive index.[5] Similarly, the destructive interference node at the surface of a reflective substrate also results in low exposure energy at the surface, which can leave material in the opened channels after development. To address this problem, the developed photoresist may be subjected to oxygen plasma to clear the channels. This process is sometimes referred to as a descumming step.

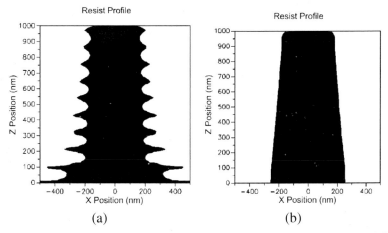

Figure 7.7 Prolith simulation of profile of a 0.5-μm photoresist feature in 1.0 μm of photoresist on a reflective substrate with (a) no postexposure bake and with (b) a postexposure bake.

After development, an additional baking step (referred to as the postbake or hardbake step) is sometimes used to drive out the remaining solvents in the resist. This procedure, while not required, can be useful if the patterned photoresist is to be used as the diffractive optical element (see the "Diffractive optics in photoresist" side bar), or to further harden the resist layer before substrate etching. All of the various parameters that have been discussed (resist type and thickness, exposure energy, bake times, process temperatures, developer type and developing time, and many other factors) affect the fidelity of the resulting lithographic image. For example, while exposed resist dissolves much faster than unexposed areas, the unirradiated areas will also begin to lose photoresist if the development time is too long. Over- or underdevelopment or exposure will decrease the fidelity of the image transfer. These concepts are illustrated in Fig. 7.8. Software packages such as

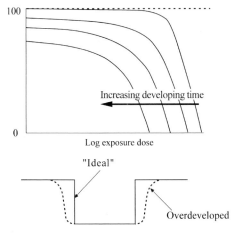

Figure 7.8 Effects of overexposure and overdevelopment of the fidelity of lithographically patterned structures.

Proxlith, Prolith, Optolith, SOLID-C, and SAMPLE are designed to simulate the exposure and development cycles. They are very useful for finding optimum sets of processing parameters.

Diffractive Optics in Photoresists

There are times when a rugged component is not required. All that is required is a DOE that will prove a design point or advance a step in a complicated development process. In such cases, these objectives can be achieved with a DOE fabricated only in photoresist, without a subsequent etching step.

As an example, consider a binary phase Dammann grating in photoresist to be used at the He:Ne laser wavelength (0.6328 μm). For a π-phase difference, the thickness of the photoresist is calculated from Eq. (6.2a) using the refractive indices of the photoresist and the surrounding material (air in this case).

If Shipley 1813 photoresist is used, the refractive index is 1.65 at the He:Ne laser wavelength (see Fig. 7.9). Therefore, the photoresist thickness needed to produce a π-phase shift should be $0.6328/(2 \times 0.65) = 0.49$ μm. However, this photoresist will not achieve this thickness, because, as is evident from Fig. 7.4, the smallest thickness that can be spun is greater than 1 μm. A photoresist with a lower viscosity is needed. For example, referring back to Fig. 7.4, an S1806 photoresist, assuming no increase in index for this type, can be used if it is spun above 6000 rpm. An alternative strategy would be to seek a photoresist with a lower refractive index.

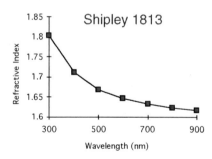

Figure 7.9 Refractive index curve for a typical photoresist.

Once the profiles have been achieved in photoresist, the piece is placed in an oven and hard baked at an elevated temperature to increase its ruggedness. This profile could also be committed to a more permanent state by using it as a master for nickel plating. The nickel shim could then be used to replicate the profile many times (see Chapter 8). In this case, the profile height must be calculated for the refractive index of the replication material.

7.1.4 Etching

In general, photoresists do not possess the appropriate optical or physical properties for most DOEs. In these cases, it is desirable to form the diffractive structures
in optical materials such as SiO_2, Si, and others that were listed in Table 6.1. These
substrate materials are typically much more rugged than photoresist because they
are more resistant to abrasion, chemicals, and thermal effects. The choice of material can also be driven by optical properties, mechanical properties, and the maturity
of the processing technology for that material. For example, SiO_2 is the material
of choice for operation from the ultraviolet through the near-infrared owing to its
transmission properties and low coefficient of thermal expansion.

Diffractive microstructures are usually fabricated in substrate materials with
dry etching techniques, although wet etching can also be acceptable for low-
resolution elements. In most cases of dry (and wet) etching, a photoresist pattern
serves as a "stencil" that protects the area underneath the photoresist during etching. As a result, only the areas *not* covered by the photoresist are removed during
the etching process, as shown in Fig. 7.10.

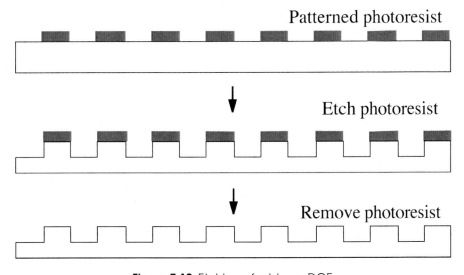

Figure 7.10 Etching of a binary DOE.

Dry etching techniques are desirable for the fabrication of DOEs for several
reasons. First, the techniques used are highly controlled and repeatable and therefore are well suited for volume manufacturing. Second, and perhaps more important, dry etching techniques are typically much more anisotropic than wet etching
methods. That is, these techniques etch preferentially in a chosen direction, usually
perpendicular to the surface of the substrate. In contrast, most wet etching techniques are isotropic (etching at equal rates in all directions), which is undesirable
for most DOEs, particularly for applications requiring very small features. Some
wet etch processes (such as for silicon) allow anisotropic etching along the crys-
talline planes of the material, but generalized DOEs cannot be made using these
processes.

Although the photoresist pattern protects the substrate from etching, the photoresist can also be etched in the process. So an important factor in etching of DOEs is the *etch selectivity*. This is defined as the etch rate of the substrate material divided by the etch rate of the masking material. A high etch selectivity is desirable for fabrication of diffractive structures with sharply defined, vertical features. It is critical that the etch mask not be completely removed before the target depth is reached in the substrate. However, thicker resist layers typically have lower resolution than thinner layers. Since thinner masks provide higher resolution imaging, etch selectivities of 2 to 3 (or greater) are desirable for etching of small features with aspect ratios greater than one. Standard diffractive etch chemistries for fused silica (usually mixtures of CF_4 and O_2 or CHF_3 and O_2) that provide etch selectivities in this range of 2 or higher are strongly preferred. Much higher etch selectivities have been achieved using photoresist as a mask with different etching parameters, particularly for etching of silicon.[8,9] Some fabrication processes for diffractive optics, such as direct writing and gray-scale lithography, benefit from etch procedures in which the selectivity has a value of near one. These procedures are discussed in more detail in Chapter 8.

As briefly mentioned in Chapter 6, it is also possible to use materials other than photoresist as the etch mask to achieve higher etch selectivities (and therefore structures with higher aspect ratios). For example, patterned chrome or aluminum layers are sometimes used as etch masks for diffractive structures. With appropriate choices of masking materials and etch parameters, extremely high etch selectivities can be achieved.[7,9,10] For example, chlorine-based etch chemistries readily etch silicon but have negligible reactions with SiO_2. As a result, a thin, patterned layer of SiO_2 can provide very high etch selectivity for chlorine-based etching of silicon.

Dry etching methods include both kinetic and reactive processes. In kinetic processes, the sample is bombarded with energetic ions that physically knock, or "sputter," the substrate molecules out of the solid matrix. Ion milling is a purely kinetic etching process that uses chemically neutral gases such as argon or krypton to etch essentially any solid material. In reactive processes, a plasma is formed from a gas or gases that chemically react with the etch material. The chemical constituents in the plasma react with the molecules in the material being etched, forming volatile gases that are removed by the etching tool. Examples of reactive etch gases include oxygen for photoresist or other organic materials, fluorine for etching fused silica, and fluorine or chlorine for etching silicon. In many cases, multiple reactive gases are used in the same etching process to enhance etch rates or achieve other effects.

The best etching processes for DOEs combine the features of both reactive and kinetic etching. In these processes, the reactive ions are accelerated vertically toward the substrate by a bias voltage, providing a strong degree of directionality to the etch process (Fig. 7.11). The directionality of the ions enables highly anisotropic etching, while the reactive nature of the ions increases the etch rate of the substrate material, thus increasing the etch selectivity of the process. Processes combining both reactive and kinetic etching include reactive ion etching, reactive

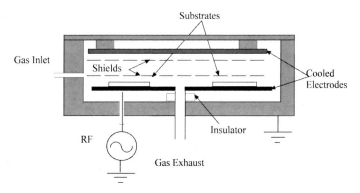

Figure 7.11 Parallel plate geometry for reactive ion etching.

ion beam etching (RIBE), and chemically assisted ion beam etching (CAIBE). Etch rates, selectivity, and quality are highly dependent on several factors, including the material being etched, the choice and mixture of etching gas, the accelerating voltage, radio frequency (RF) power, pressure, and other factors. Typical etch rates for fused silica range from about 15 to 50 nm/min, but rates approaching 1000 nm/min are achievable. Similarly, rates for silicon etching usually range from about 15 to 3000 nm/min, but rates over 10 μm/min have been demonstrated.

Fused silica DOEs can be also be wet etched in buffered hydrofluoric acid, with typical etch rates of ∼100 nm/min. This approach has several advantages, including the fact that it is very low cost and thus easily accessible. Etched areas are extremely smooth, and the process is highly repeatable if appropriate care is taken. Disadvantages include the fact that the process is extremely dangerous and special precautions must be taken while working with hydrofluoric acid. Also, the isotropic nature of the etch causes undercutting and sloped side walls, and is thus inappropriate for fine patterns. The etch rate is also temperature dependent, so precautions must be taken to achieve good process repeatability. This rapid etch rate is advantageous in that it shortens the time needed to fabricate a given diffractive structure. The rate can also be a disadvantage because accurate realization of a specific etch depth requires tighter controls on etch time than a slower etch process.

7.2 Binary Optics

As discussed earlier, one of the most common fabrication methods for diffractive optics uses the photolithographic fabrication sequences and equipment that we have discussed. Diffractive optics fabricated using binary lithographic masks were first made in the early 1970s.[10,11] Efficient manufacturing techniques using binary lithographic masks were demonstrated in the mid-1980s by researchers at MIT's Lincoln Laboratory.[12,13] Unlike earlier lithographic masking approaches that generated $N + 1$ phase steps from N masks, this approach generated diffractive structures with up to 2^N phase levels from N masks. DOEs fabricated using this approach are commonly referred to as binary optics.[14]

The first mask is used to lithographically expose a photoresist layer on the substrate. The patterned substrate is developed and etched, with the patterned photoresist protecting portions of the substrate. A diffractive element with many levels is fabricated by using multiple masks and repeating the lithographic processing steps. This process is illustrated schematically in Fig. 7.12. While we show the fabrication of the largest features first for illustration purposes, it should be noted that in practice it is often desirable to fabricate the smallest features first so that the most critical lithography steps are performed on planar surfaces for the highest patterning fidelity. Photoresist coatings on nonplanar surfaces are less uniform than coatings on planar surfaces, resulting in additional process variability and reduced fidelity. Furthermore, it is more difficult to simultaneously pattern "small" features at different heights (i.e., at the top of the substrate surface and the bottom of a trench already patterned in the substrate) with optimum fidelity owing to diffractive spreading of the light passing through the mask (for contact and proximity lithography) or limitations on the depth of focus (for projection lithography). For these reasons, it is generally best to pattern the largest features last so that the effects of these various factors are minimized.[15] It is also possible to build up surface relief structures for diffractive optics by thin-film deposition using similar techniques.[16,17]

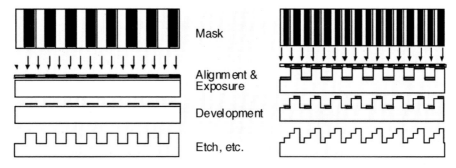

Figure 7.12 Binary mask 2^N fabrication process for DOEs.

As discussed in Chapter 6, the first step of the process consists of decomposing the diffractive design into a set of electron beam-written masks. The mask encoding method (and thus the locations of the transition points on the masks) is dependent on how the final phase structure will be fabricated. A sample of a three-mask set for an eight-phase level diffractive beam shaper using the binary optics approach is shown in Fig. 7.13, along with the resulting surface structures that are obtained by sequential patterning and etching. Scanning electron microscope images of sample multilevel diffractive structures fabricated using the binary optics approach are shown in Fig. 7.14.

Because of their ties to semiconductor manufacturing, binary optics techniques are probably the most mature of the lithographic fabrication technologies. The fabrication procedures are well understood, characterized, and repeatable. Like microchips, DOEs made with this method are fabricated at the wafer level, providing great economies of scale for volume production. Very small feature sizes

Figure 7.13 (Top) Binary masks and (bottom) resulting surface patterns from sequential processing with these masks for an eight-phase level diffractive beam shaper.

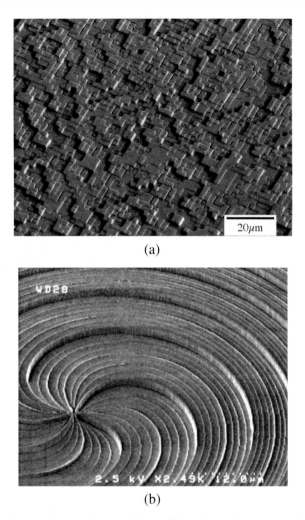

(a)

(b)

Figure 7.14 SEM images of eight-phase level diffractive elements. (a) Spot array generator. (b) Spiral beam-shaping element. (Both photos courtesy of Digital Optics Corp., Charlotte, NC.)

can be achieved using both contact printing (~0.5 µm) and projection lithography (<0.3 µm). For most elements, the use of a relatively small number of lithographic masks provides a good fit to the ideal surface shape and excellent optical performance. As discussed in Chapter 2, an eight-phase level (three masks) diffractive lens has a scalar theoretical efficiency of 95%, and a sixteen-level (four masks) lens has a 99% theoretical efficiency. The biggest disadvantage is the need for multiple processing steps to generate multilevel diffractives. This increases production costs over single-step procedures. However, the mature nature of the fabrication technology enables high production yields, which ultimately reduces costs. Although the need for multiple alignments can introduce level-to-level mask alignment errors that decrease element performance, advanced manufacturing methods can reduce typical alignment errors to dimensions commensurate with the size of the address units in the electron beam-written masks (≤ 0.1 μm).

7.3 Conclusion

The principles of photolithography provide the basis for fabrication of diffractive optical elements. Along with the binary optics fabrication methods just discussed, there are other techniques that make use of photolithographic principles to generate DOEs. In the next chapter we survey these other fabrication techniques.

References

1. S. Wolf and R.N. Tauber, *Silicon Processing for the VLSI Era, Vol. 1: Process Technology*. Lattice Press, Sunset Beach, CA (1986).
2. D.J. Elliot, *Microlithography Process Technology for IC Fabrication*. McGraw-Hill, New York (1986).
3. P. Rai-Choudhury, *Handbook of Microlithography, Micromachining, and Microfabrication, Vol. 1: Microlithography*. SPIE Press, Bellingham, WA (1997).
4. J.R. Sheats and B.W. Smith, *Microlithography Science and Technology*. Marcel Dekker, New York (1998).
5. H.J. Levinson, *Principles of Lithography*. SPIE Press, Bellingham, WA (2001).
6. T.J. Suleski, "Diffractive optics fabrication," in *Encyclopedia of Optical Engineering*, R.G. Driggers, Ed. Marcel Dekker, New York, pp. 374–387 (2003).
7. M.J. Madou, *Fundamentals of Microfabrication: The Science of Miniaturization*, 2nd ed. CRC Press LLC, Boca Raton, FL (2002).
8. L. Laermer and A. Schlip, "Method of anisotropically etching silicon," Robert Bosch GmbH, U.S. Patent 5,501,893.
9. W.-C. Tian, J.W. Weigold, and S.W. Pang, "Comparison of Cl_2 and F-based dry etching for high aspect ratio Si microstructures etched with an inductively coupled plasma source," *J. Vac. Sci. Tech. B.* **18**, pp. 1890–1896 (2000).
10. L. d'Auria, J.P. Huignard, A.M. Roy, and E. Spitz, "Photolithographic fabrication of thin film lenses," *Opt. Comm.* **5**, pp. 232–235 (1972).
11. J.J. Clair and C.I. Abitbol, "Recent advances in phase profile generation," in *Progress in Optics*, Vol. XVI, E. Wolf, Ed. North Holland, Amsterdam, pp. 73–117 (1978).
12. W.B. Veldkamp and G.J. Swanson, "Developments in fabrication of binary optics," in *Intl. Conference on Computer Generated Holography*, S.H. Lee, Ed. *Proc. SPIE* **437**, pp. 54–59 (1983).
13. G.J. Swanson, "Binary optics technology: the theory and design of multi-level diffractive optical elements," MIT Lincoln Laboratory Report 854. Massachusetts Institute of Technology, Cambridge, MA (1989).
14. M.B. Stern, "Binary optics fabrication," in *Micro-optics: Elements, Systems, and Applications*, H.P. Herzig, Ed. Taylor and Francis, London, pp. 53–85 (1997).

15. J.M. Miller, M.R. Taghizadeh, J. Turunen, and N. Ross, "Multilevel-grating array generators: fabrication error analysis and experiments," *Appl. Opt.* **32**, pp. 2519–2525 (1993).

16. J. Jahns and S.J. Walker, "Two-dimensional array of diffractive microlenses fabricated by thin-film deposition," *Appl. Opt.* **29**, pp. 931 (1990).

17. E. Pawlowski and B. Kuhlow, "Antireflection-coated diffractive optical elements fabricated by thin-film deposition," *Opt. Eng.* **33**, pp. 3537–3546 (1994).

Chapter 8

Survey of Fabrication Techniques for Diffractive Optical Elements

Fabrication of diffractive optical elements with binary transmission lithographic masks was used in Chapters 6 and 7 to demonstrate a number of principles. However, there are other techniques that have been developed for fabrication of diffractive microstructures. As previously noted, these techniques fall into three main categories: lithographic methods, direct machining, and replication. In this chapter we present a survey of other techniques used to make diffractive optics. Additional overviews of the history and techniques of modern micro-optics fabrication can be found in several references.[1–7]

8.1 Lithographic Techniques

In addition to the binary mask method discussed in Chapter 7, other lithographic techniques are commonly used for fabrication of surface relief structures. These techniques differ from those used in creating binary optics primarily in the method used to expose the pattern on the photoresist. Some of the methods use a different etching procedure as well. We now consider several additional lithographic techniques for fabrication of diffractive optical elements, including direct writing, interferometric exposure, gray-scale lithography, and near-field holography (NFH).

8.1.1 Direct writing

One common approach to the fabrication of micro-optical elements is to write the exposure pattern directly into the photoresist layer using an electron beam or laser beam, as shown in Fig. 8.1(a).[8–13] Instead of establishing the pattern through a series of mask exposures, the intensity of the beam is varied so that the local exposure is proportional to the required depth of the resist. To ensure that the features are reliably produced, the spot sizes for e-beam direct-write systems are usually ~0.1 μm or larger. Laser direct-write systems typically use tightly focused beams

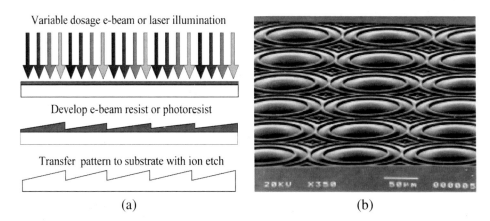

Variable dosage e-beam or laser illumination

Develop e-beam resist or photoresist

Transfer pattern to substrate with ion etch

(a)   (b)

Figure 8.1 (a) Direct-write fabrication process for DOEs. (b) Diffractive lens array fabricated using laser direct writing. (Courtesy of RPC Photonics, Inc., Rochester, NY.)

from helium-cadmium (He:Cd) or argon-ion lasers to generate spot sizes from ~ 1 to 5 μm.

As discussed in Chapter 7 and illustrated in Fig. 7.6, the amount of photoresist remaining after development for a given exposure energy is determined by the photoresist response. If this relationship is understood, it is possible to map an exposure dosage to a specific topographical depth after development of the photoresist. When the resist is developed, the depth of the local surface relief structure in the resist layer is proportional to the dosage delivered to that area by the e-beam or laser source. The photoresist response curve is typically nonlinear, so it is sometimes useful to work in the "linear" region beyond the knee of the resist response curve. Using this approach, it is possible to create essentially continuous surface relief structures by varying the power level or dwell time of the beam as it is scanned across the substrate. After development, the surface shape can be transferred directly into the underlying substrate using dry etching. An example of a diffractive lens array fabricated using laser direct writing is shown in Fig. 8.1(b).

The use of direct-write systems for fabrication of DOEs offers many potential advantages. Because these processes eliminate the need for lithographic masks, the time and expense of creating prototype elements are reduced. If the dosage delivered to a given area is precisely controlled as the beam is scanned across the substrate, a large number of phase levels (256 or more) can be generated. One disadvantage of direct writing is that it is a serial process. That is, each element must be written one at a time by the scanning beam. Because of the serial nature of the direct-write process, it is inappropriate for volume manufacturing of DOEs unless it is coupled with the replication techniques discussed later. In many cases, the performance advantage of more phase levels is offset by changes in surface profiles caused by variations in lithographic chemistry. In addition, the finite writing-spot sizes cause nonvertical side walls. Also, accurate transfer etching of continuous-relief structures into the substrate requires a highly anisotropic etch mechanism and a predictable etch selectivity S. If $S = 1$, then the transferred pattern is essentially identical to the photoresist structure (assuming that the etch is purely anisotropic). Similarly, an etch with $S < 1$ results in a shallower structure and $S > 1$ results in a deeper structure. As discussed briefly in Chapter 7, etch selectivity is dependent on multiple factors. As an example, for reactive ion etching of shapes into fused silica, the relative concentrations of etch gases such as oxygen and CHF_3 (or another fluorine-containing etch gas) can be changed. In this case, higher concentrations of oxygen cause the photoresist layer to etch faster, resulting in a lower selectivity. Changes in the photoresist area relative to the substrate area during the transfer etch can generate loading effects that cause continuous changes in etch selectivity during the etch, distorting the shape of the relief structure and reducing the optical performance.

8.1.2 Interferometric exposure

A discussion of the broad range of techniques available for creation of holographic optical elements is beyond the scope of this text. However, the primary method for the fabrication of diffractive phase gratings with extremely small periods is based on holographic interference. This maskless process creates interference patterns used to photolithographically expose photoresist, as shown in Fig. 8.2. The resist layer is then developed and the pattern etched into the substrate as previously described. This general approach has been used to create diffraction gratings with periods of less than 0.3 μm for multiple applications, including antireflective surfaces and wave plates.[14]

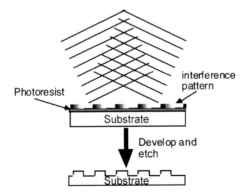

Figure 8.2 Interferometric fabrication of high-frequency diffraction gratings.

8.1.3 Gray-scale lithography

Another fabrication method for DOEs that has a great deal of promise is referred to as gray-scale lithography. This manufacturing technique takes advantage of the response of photoresist to varying exposures to allow the fabrication of multilevel DOEs with a single lithographic masking and etching step. Fabrication of micro-optical components using gray-scale lithography has been reported in the literature for some time,[15-17] but has only recently made inroads into commercial manufacturing of micro-optics.

The principles of gray-scale lithography are straightforward. In this process (illustrated in Fig. 8.3), a lithographic mask with a spatially varying transmission profile is imaged onto a photoresist layer, either through contact or projection lithography. As with direct writing, the local surface relief depth in photoresist is proportional to the energy transmitted through that area of the gray-scale mask after development (see Fig. 8.3). The resulting surface relief pattern can then be transferred into the substrate through controlled plasma etching or replicated using casting techniques. Examples of DOEs manufactured in photoresist and fused silica using the gray-scale lithographic process are shown in Fig. 8.4 and Fig. 8.5.

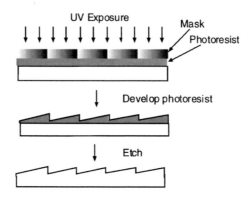

Figure 8.3 Gray-scale lithographic technique for DOE fabrication.

The benefits of gray-scale lithography are readily apparent. Like direct-write techniques, it allows the fabrication of microstructures with an essentially arbitrary number of phase levels. Like the binary optics approach, it allows the fabrication of multiple components at the wafer level. Furthermore, as a single-step lithographic

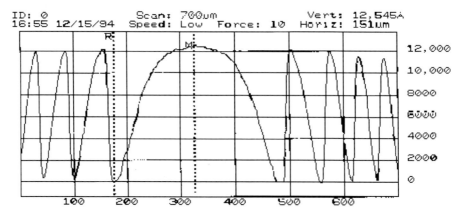

Figure 8.4 Diffractive lens fabricated in photoresist using gray-scale lithography and methods of desktop publishing (see "PostScript methods of mask fabrication" in Chapter 6).

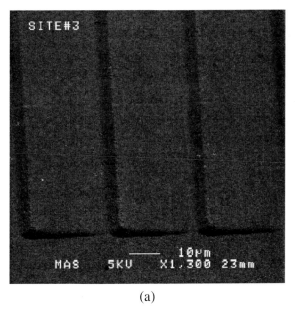

(a)

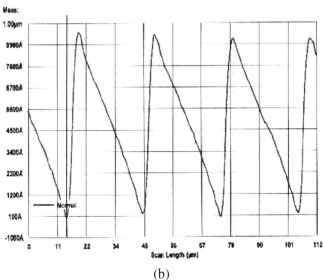

(b)

Figure 8.5 (a) SEM measurement and (b) profilometer trace of 30-µm period blazed gratings in fused silica fabricated using gray-scale lithography. (Courtesy of Digital Optics Corp., Charlotte, NC.)

procedure, it eliminates the need for multiple mask alignments. The elimination of processing steps reduces fabrication costs and reduces the possibility of fabrication errors resulting from mask misalignment. Thus this method has the potential to increase device performance and reduce manufacturing costs.

Although gray-scale lithography offers potential performance and cost benefits, it presents several significant challenges as well. First, the masks are currently more difficult to obtain and may cost significantly more than a set of binary masks. Second, the gray-scale lithographic process is much more sensitive to variations in lithographic materials and process parameters than binary techniques. The potential cost benefits gained by reducing the number of processing steps may be diminished by lower yields. Like optical components fabricated using direct-write techniques, the potential performance improvements gained from using a larger number of phase levels may be offset by changes in the surface shape resulting from variations in lithographic parameters, nonvertical side walls, and deviations in selectivity during the transfer etch. Still, the potential benefits of the technology make it a worthwhile addition to a microfabrication engineer's toolkit.

8.1.4 Near-field holography

Near-field holography (NFH) is a promising technique for fabrication of high-spatial frequency diffractive gratings for a variety of applications, including waveguide couplers, subwavelength antireflection structures, wavelength division multiplexing, and many others. The dimensions of these gratings typically require that each grating be fabricated through direct electron-beam writing or through conventional interferometric fabrication. However, both of these methods have significant drawbacks. As noted earlier, direct e-beam writing is a serial process that is both time-consuming and expensive, while interferometric exposure offers relatively limited flexibility for fabricating different structures. NFH has the potential to provide a low-cost method of producing these high-frequency gratings.

NFH is similar to the interferometric lithographic process previously described except that NFH uses near-field diffraction patterns from diffractive phase masks for photoresist exposure.[18,19] One method, illustrated in Fig. 8.6, uses a phase grating that minimizes the zeroth-order energy transmitted with normally incident light.[20] The near-field irradiance pattern generated from interference of the +1 and −1 transmitted diffraction orders has a period one-half of that of the original phase mask, which is beyond the realm of conventional contact lithography. This approach has also been applied to the manufacture of linear, rectilinear, and circular geometries.[21] An example of a diffractive phase mask and a grating made from it using NFH are shown in Fig. 8.7.

8.1.5 Refractive micro-optics

Although the focus of this text is diffractive optics, a discussion of refractive micro-optics is included for two reasons. First, the elements are fabricated using many of the same methods as diffractive optical elements. Second, they are commonly used for many micro-optical applications, along with diffractive optics. The heights of diffractive structures in the visible and near-infrared are usually no more than 3 to 4 μm. In comparison, refractive microstructures are commonly fabricated with heights of many tens of microns and with diameters of up to several hundred microns or more.

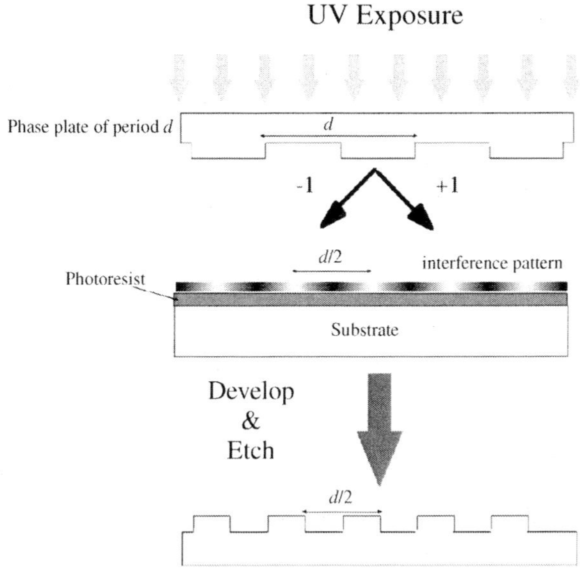

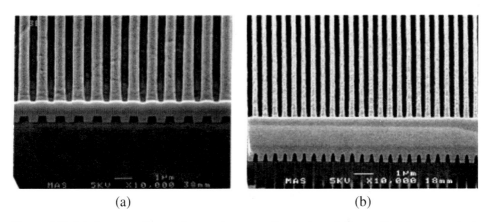

Figure 8.6 Manufacturing technique for high-frequency diffraction gratings using near-field holography. (Courtesy of Digital Optics Corp., Charlotte, NC.)

Figure 8.7 (a) Cross section of linear phase grating in fused silica with 1.0-μm period. (b) Cross section of linear phase grating in fused silica with 0.5-μm period manufactured from the phase mask shown using the near-field holographic technique showing Fig. 8.6. (Courtesy of Digital Optics Corp., Charlotte, NC.)

A primary method for the fabrication of refractive microlenses uses a binary transmission mask to pattern a cylinder of photoresist after exposure and development, as discussed in Chapter 7. The photoresist is then melted, or "reflowed" on a hot plate or in an oven.[22] Surface tension forms a high-quality refractive microlens through this process when the ratio of the lens diameter to the lens height is in an appropriate range. A broader range of refractive surface shapes can be patterned in photoresist using gray-scale lithography or direct-writing techniques (usually laser direct writing). All of these methods require precision control of etching parameters to accurately transfer the photoresist shape into the underlying substrate. Different etch selectivities can be also be used to change the lens shape during transfer etching. The "reflow" method of refractive microlens fabrication and a sample lens are illustrated in Fig. 8.8.

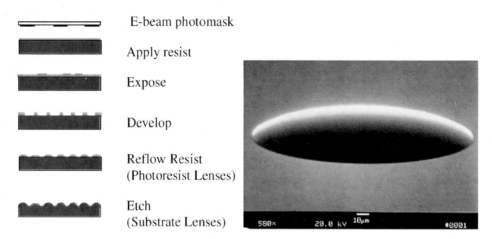

E-beam photomask

Apply resist

Expose

Develop

Reflow Resist
(Photoresist Lenses)

Etch
(Substrate Lenses)

Figure 8.8 (Left) Reflow method for fabrication of refractive microlenses. (Right) SEM image of refractive microlens in fused silica. (Courtesy of Digital Optics Corp., Charlotte, NC.)

Refractive micro-optics generally have a higher efficiency and operate across a much wider bandwidth than DOEs. Their disadvantages include the difficulty in making aspherical lens shapes and size limitations arising from the fabrication methods. In particular, the time needed to transfer the photoresist lens shape into the underlying substrate can be significant. For example, a refractive microlens with a 40-μm height takes more then 33 h to etch into fused silica at 200 Å/min, a typical etch rate for reactive ion etching of this material. Much faster etch rates (and thus shorter etching times) can be achieved with other plasma etching systems, including inductively coupled plasma (ICP) etchers, but the requirements for precise process control remain unchanged.

8.2 Direct Machining

Unlike photolithographic techniques, direct machining methods for DOEs use no intermediate processing steps to make microstructures. The structures are formed through direct removal of the optical material. These processes can produce high-quality gratings, but they are relatively slow. As a consequence, they are commonly used to make "master" gratings for the replication processes discussed in Sec. 8.3.

8.2.1 Mechanical ruling

Since the earliest work performed by Fraunhofer and Rowland, significant advances have been made in mechanical ruling techniques. Based on computer con-

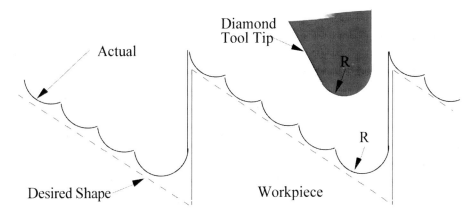

Figure 8.9 Mechanical ruling of a blazed grating structure using a single-point diamond stylus. (Courtesy of Carmiña Londoño[26], Polaroid.)

trol, modern techniques are used to fabricate blazed diffraction gratings of great quality with thousands of grooves per millimeter.[23,24] These mechanical ruling methods use a sharp stylus tip to scrape away the substrate material in a highly controlled manner. Owing to the finite size of the tip, the tool cannot cut a perfectly smooth profile, so it is necessary to achieve a balance between the fineness of the cut and accuracy of the surface profile, as shown in Fig. 8.9. However, in some cases an entire blazed grating period can be cut at one time with the edge of a specially machined, triangular stylus.[25]

8.2.2 Diamond turning

Diffractive microstructures can also be generated by rotating the substrate as it is brought into contact with the stylus using a technique commonly referred to as diamond turning. Single-point diamond turning (SPDT) is a technique that has been used for many years for the fabrication of refractive optics[27] and then for DOEs.[26,28–30] This technique requires precise mechanical control and is limited to radially symmetric patterns. A sample setup for diamond turning is illustrated in Fig. 8.10.

As noted earlier, it is necessary to achieve a balance between fineness of cut and surface profile accuracy because of the finite size of the stylus tip. Again, a specially machined triangular stylus can be used to generate an entire lens zone with a single cut. This technique allows the fabrication of more precise approximations to the blazed zone shape, but the angle of the stylus tip must be rotated in order to generate blazed zones of different widths, as is required for a diffractive lens.

Direct machining by ruling engines or diamond turning limits the structures to either straight line or circularly symmetrical gratings, respectively. It would be extremely difficult (if not impossible) to fabricate a generalized two-dimensional spot array generator, for example, using either of these methods. However, recent

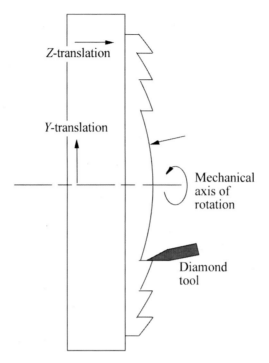

Figure 8.10 Single-point diamond turning of a diffractive lens.

advances in multiaxis micromachining and microgrinding systems may provide additional flexibility for fabrication of micro-optics.[31,32]

8.2.3 Other methods of direct machining

Additional direct machining methods have been developed for fabrication of diffractive structures in which optical material is directly removed by energetic beams of ions or energy. In one method, referred to as focused ion beam (FIB) milling, a focused beam of ions sputters the atoms in the material off of the surface.[33,34] In another method, referred to as laser ablation, a focused beam from an excimer laser is used to directly machine the surface.[35,36] These procedures are similar to the direct-writing techniques described earlier because the processes are serial in nature and the local depth of the diffractive structure is proportional to the length of time that the beam dwells on a specific location. The distinction between these two methods is that direct-writing techniques form microstructures in developed photoresist, whereas focused ion beam milling and laser ablation create the structures directly in the chosen optical material. The laser ablation technique has also been used to machine complete patterns at one time by imaging a diffractive mask pattern with an objective lens and an excimer laser.[37] Unlike mechanical ruling and diamond turning, these processes can be used to create essentially arbitrary surfaces. The resolution of the FIB process is comparable to that of electron-beam direct writing, while the resolution of laser ablation is roughly comparable to laser direct writing.

8.3 Replication

Because of the time and cost required to fabricate diffractive optics by direct machining and lithography techniques, it is desirable to make precise copies of these surface relief structures in other materials at low cost.[38,39] This can be done by making a "master" diffractive element using either direct machining or lithographic techniques. In many cases, a metal copy, or "shim," of the original microstructure is electroformed in robust metals such as nickel [Fig. 8.11(a)]. The master element or shim is then brought into contact with a formable material, usually a curable polymer or thermoplastic, using one of several methods, such as molding, embossing, or casting [Figs. 8.11(b) to 8.11(d)]. After the plastic has been solidified, the replica is separated from the mold. It is usually necessary to treat the mold structure with a "release agent" so that the copy releases easily from the mold.

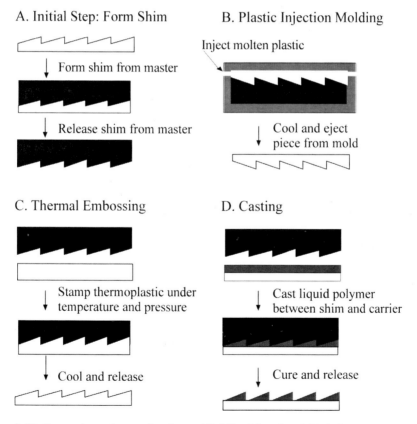

Figure 8.11 Procedures for replication of DOEs. After the initial step that forms the shim from the master profile, the DOE may be replicated by one of three techniques.

8.3.1 Plastic injection molding

Plastic injection molding is used for fabrication of a vast range of products in modern society, from compact disks to toys and plastic cups. In this process, a metal mold cavity is fabricated through various machining methods. For diffractive structures, a "mold insert" containing the diffractive pattern is usually mounted in the mold cavity. Molten plastic is injected into the mold cavity, then the plastic part is "ejected" from the mold after the plastic cools and hardens [Fig. 8.11(b)]. Two typical plastics for plastic injection molding of diffractive optics are polycarbonate (PC) and polymethyl methacrylate (PMMA). The optical quality of a diffractive optic fabricated in this manner is generally lower than the master element. Because it is capable of producing literally millions of parts per month at low cost, plastic injection molding is appropriate for volume manufacturing of DOEs that do not require excessively tight fabrication tolerances.

8.3.2 Thermal embossing

Thermal embossing of diffractive optics is performed by bringing the shim into contact with a solid plastic under heat and pressure. As the temperature of the plastic is raised above the glass transition temperature, T_g, the plastic softens and conforms to the shape of the shim. Once the stack is cooled and the pressure is removed, the embossed plastic replica is separated from the shim. This basic procedure can be performed with a single shim through stamping, or with a shim mounted on a roller arrangement so that the diffractive surface structures can be continuously embossed onto sheets of thermoplastic foil. Thermal roller embossing is the primary method for producing security holograms, such as those found on most credit cards. Like plastic injection molding, thermal embossing can be used to produce large quantities of diffractive structures of reasonable quality and low cost.

8.3.3 Casting and UV embossing

Casting methods for replication of diffractive optics are similar to thermal embossing techniques, with the key distinction that the plastic material is already in liquid form when it is brought into contact with the shim. Using casting methods, a liquid polymer layer is sandwiched between a blank substrate and the mold surface. The polymer layer is then solidified, or "cured," either by exposure to UV light for photopolymers, or through baking for thermoset polymers. The mold and the substrate are separated after the polymer layer is cured. This technique is used to form high-quality replicas. Like thermal embossing, this basic technique can also be applied to continuous roller replication into UV-curable materials on plastic sheets. An approach that combines the high pressure of thermal embossing with principles of UV embossing has also been demonstrated for dry photopolymers.[38] Similarly, "sol-gel" materials that reduce to SiO_2 after thermal curing have been used to replicate glass micro-optics onto substrates, although great care must be taken with this process because the material can undergo significant shrinkage during curing.[40]

8.3.4 Soft lithography

A relatively new set of processing methods, collectively referred to as soft lithography, provides additional approaches to microfabrication of diffractive structures.[41] In soft lithography, high-fidelity replicas of microsurface patterns are made in elastomeric compounds (Fig. 8.12). These elastomeric "stampers" can then be used either as inexpensive replicas or as molds or stamps for additional processing.

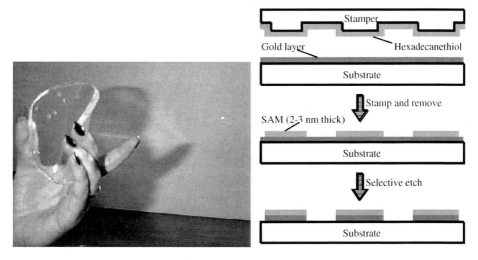

Figure 8.12 (Left) Elastomeric replica element containing diffractive optical structures. (Right) Illustration of a microcontact printing process for gold patterning. (Courtesy of Digital Optics Corp., Charlotte, NC.)

The two main processes of soft lithography are known as microcontact printing[42,43] and microtransfer molding.[44–46] Both processes are relatively straightforward. In microcontact printing, a chemical "ink" is applied to an elastomeric stamper element. The ink is transferred to a substrate by stamping a metal-coated substrate with the elastomeric element. This forms a self-assembled monolayer (SAM) pattern in areas of contact. The patterned monolayer on the substrate then serves as a selective etch mask, as shown in Fig. 8.12. The choice of chemical ink is determined by the material to be patterned. This process is typically used to fabricate patterns in metals, although the microcontact printing method has also been used as part of a more complex process to fabricate microstructures in fused silica.[47] The second process, microtransfer molding, is used to replicate microstructures in polymers and other materials. This process is similar to the casting methods of replication illustrated in Fig. 8.11(d), except that the mold "shim" is made from an elastomeric material. After the molded polymer is cured, the optical function can be left in polymer form on the substrate surface, or the surface shape can be

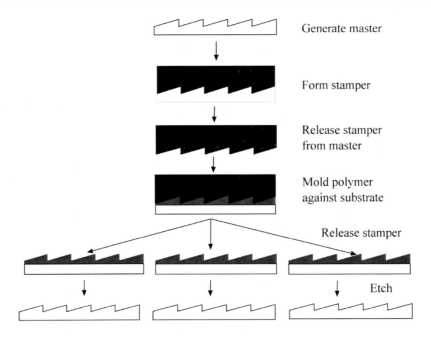

Generate master

Form stamper

Release stamper
from master

Mold polymer
against substrate

Release stamper

Etch

Figure 8.13 Microtransfer molding fabrication process for micro-optics.

transferred into the underlying substrate in a manner analogous to the etching procedures for gray-scale lithography or direct writing.[46] This process is illustrated in Fig. 8.13.

Although it is still necessary to manufacture the master grating using conventional methods, soft lithography permits the inexpensive fabrication of many high-precision optical components. Soft lithography has been used to create patterns with features smaller than 100 nm. The flexible nature of the elastomeric stamper elements used for soft lithography also enables microfabrication on curved surfaces.[41] These approaches have some drawbacks, particularly for the fabrication of multilevel diffractives (microcontact printing) and diffractives requiring precise registration. Additional research is required to determine the role that these techniques will play in DOE manufacturing.

8.4 Comparison of Fabrication Methods for DOEs

A wide variety of fabrication methods for diffractive optical elements have been described in this chapter. Each of these techniques has various strengths and weaknesses. Some are better for volume manufacturing, and some for small quantities. The choice of material can limit the choice of fabrication method that can be used. Similarly, the optical functionality of the component can also drive the choice of fabrication method. Table 8.1 presents a summary of material types, strengths, and weaknesses for the various fabrication techniques.

Table 8.1 Comparison of fabrication methods for DOEs.

Lithographic Techniques			
Method	**Materials**	**Strengths**	**Weaknesses**
Binary optics	Photoresist Glasses Semiconductors	Volume production Surface precision Small features Very flexible	Prototyping costs Multiple processing steps for high efficiency
Grayscale lithography	Photoresist Glasses Semiconductors	Volume production Analog surfaces Very flexible Mask costs	Process variability Surface precision
Direct writing (Laser)	Photoresist Glasses Semiconductors	Analog surfaces Rapid prototyping Very flexible	Serial process Low volumes Edge definition
Direct writing (E-beam)	Photoresist Glasses Semiconductors	Very small features Very flexible Analog surfaces possible	Slow serial process Equipment costs
Interferometric exposure, Near-field holography	Photoresist Glasses Semiconductors	Small features Volume production	Limited flexibility
Direct Machining			
Method	**Materials**	**Strengths**	**Weaknesses**
Mechanical ruling	Metals, plastics, and some glasses	1D gratings	Limited flexibility Very slow
Diamond turning	Metals, plastics, and some glasses	Lens fabrication Rapid prototyping	Limited flexibility (radial symmetry) Low volumes
Focused ion beam milling	Glasses Semiconductors Metals	Analog surfaces possible Small features Very flexible	Slow serial process Equipment costs
Laser ablation	Plastics Glasses	Rapid prototyping Very flexible	Low volumes only Edge definition
Replication			
Method	**Materials**	**Strengths**	**Weaknesses**
Injection molding	Plastics	High volumes Integrated packaging features	Relative precision Limited to plastics Expensive tooling
Thermal embossing	Plastics	High volumes	Relative precision limited to plastics
Casting	Plastics Silica (sol-gels)	High precision replicas (plastics) Glass replicas (sol-gels)	Relatively slow Sol-gel shrinkage
Soft lithography	Semiconductors, Metals, plastics, and glasses	High volumes Planar and curved surfaces	Relative precision Mask distortion

References

1. H.P. Herzig (Ed.), *Micro-optics: Elements, Systems, and Applications.* Taylor and Francis, London (1997).

2. M. Kufner and S. Kufner, *Micro-optics and Lithography*, VUB Press, Brussels (1997).

3. H.O. Sankur and M.E. Motamedi, "Microoptics development in the past decade," in *Micromachining Technology for Micro-Optics*, S.H. Lee and E.G. Johnson, Eds., *Proc. SPIE* **4179**, pp. 30–55 (2000).

4. S. Sinzinger and J. Jahns, *Microoptics*, Wiley-VCH, Weinheim, Germany (1999).

5. J. Mait, "From ink bottles to e-beams: a historical perspective of diffractive optics," in *Optical Processing and Computing: A Tribute to Adolf Lohmann*, D. Casasent, H.J. Caulfield, W.J. Dallas et al., Eds., *Proc. SPIE* **4392** (2001).

6. T.J. Suleski and R.D. Te Kolste, "A roadmap for micro-optics fabrication," in *Lithographic and Micromachining Techniques for Optical Component Fabrication*, E.B. Kley and H.P. Herzig, Eds., *Proc. SPIE* **4440**, pp. 1–15 (2001).

7. T.J. Suleski, "Diffractive optics fabrication," in *Encyclopedia of Optical Engineering*, R.G. Driggers, Ed., Marcel Dekker, New York (2003).

8. T. Fujita, H. Nishihara, and J. Koyama, "Fabrication of micro-lenses using electron beam lithography," *Opt. Lett.* **6**, pp. 613–615 (1981).

9. T. Fujita, H. Nishihara, and J. Koyama, "Blazed gratings and Fresnel lenses fabricated by electron-beam lithography," *Opt. Lett.* **7**, pp. 578–580 (1982).

10. M.T. Gale and K. Knop, "The fabrication of fine lens arrays by laser beam writing," in *Industrial Applications of Laser Technology*, W.F. Fagan, Ed., *Proc. SPIE* **398**, pp. 347–353 (1983).

11. M.T. Gale, M. Rossi, H. Schütz, P. Ehbets et al., "Continuous-relief diffractive optical elements for two-dimensional array generation," *Appl. Opt.* **32**, pp. 2526–2533 (1993).

12. J.P. Bowen, R.L. Michaels, and C.G. Blough, "Generation of large-diameter diffractive optical elements with laser pattern generation," *Appl. Opt.* **36**, pp. 8970–8975 (1997).

13. M.T. Gale, "Direct writing of continuous-relief micro-optics," in *Micro-optics: Elements, Systems, and Applications*, H.P. Herzig, Ed., Taylor and Francis, London, pp. 87–126 (1997).

14. R.C. Enger and S.K. Case, "Optical elements with ultrahigh spatial-frequency surface corrugations," *Appl. Opt.* **22** (20), pp. 3220–3228 (1983).

15. A.G. Poleshchuk, "Fabrication of relief-phase structures with continuous and multilevel profiles for diffractive optics," *Optoelectron. Instrum. Data Process* **1**, pp. 67–79 (1992).

16. T.J. Suleski and D.C. O'Shea, "Gray-scale masks for diffractive-optics fabrication: I. Commercial slide imagers," *Appl. Opt.* **34**, pp. 7507–7517 (1995).

17. W. Daschner, P. Long, M. Larsson, and S.H. Lee, "Fabrication of diffractive optical elements using a single optical exposure with a gray level mask," *J. Vac. Sci. Technol. B* **13**, pp. 2729–2731 (1995).

18. M. Okai, S. Tsuji, N. Chinone, and T. Harada, "Novel method to fabricate corrugation for a $\lambda/4$-shifted distributed feedback laser using a grating photomask," *Appl. Phys. Lett.* **55**(5), pp. 415–417 (1989).

19. D.M. Tennant, K.F. Dreyer, K. Feder, R.P. Gnall et al., "Advances in near field holographic grating mask technology," *J. Vac. Sci. Technol. B* **12**(6), pp. 3689–3694 (1994).

20. P.I. Jensen and A. Sudbo, "Bragg gratings for 1.55-µm wavelength fabricated on semiconductor material by grating-period doubling using a phase mask," *IEEE Photon. Technol. Lett.* **7**(7), pp. 783–785 (1995).

21. T.J. Suleski, B. Baggett, W.F. Delaney, C. Koehler et al., "Fabrication of high spatial frequency gratings through computer generated near-field holography," *Opt. Lett.* **24**, pp. 602–604 (1999).

22. Z.D. Popovic, R.A. Sprague, and G.A. Neville Connell, "Technique for monolithic fabrication of microlens arrays," *Appl. Opt.* **27**, pp. 1281–1284 (1988).

23. C. Palmer, *Diffraction Grating Handbook*. Richardson Grating Laboratory, Rochester, NY (2002). Available as electonically at www.gratinglab.com/Library/hanbook/handbook.asp

24. "Diffraction Gratings," product information, Richardson Grating Laboratory, 5th ed. 705 St. Paul St., Rochester, NY, 14605 (2003).

25. M.B. Fleming and M.C. Hutley, "Blazed diffractive optics," *Appl. Opt.* **36**, pp. 4635–4643 (1997).

26. C. Londoño, "Design and fabrication of surface relief diffractive optical elements, or kinoforms, with examples for optical athermalization," Ph.D. thesis, Tufts University, Medford, MA (1992).

27. T.T. Saito, "Diamond turning of optics: The past, the present, and the exciting future," *Opt. Eng.* **17**, pp. 570–573 (1978).

28. M.J. Riedl and J.T. McCann, "Analysis and performance limits of diamond-turned diffractive lenses for the 3-5 and 8-12 micrometer regions," in *Infrared Optical Design and Fabrication*, R. Harsman and W.J. Smith, Eds., *SPIE Press* **CR38**, pp. 153–163 (1991).

29. P.P. Clark and C. Londoño, "Production of kinoforms by single-point diamond machining," *Opt. News* (12), pp. 39–40 (1989).

30. C.G. Blough, M. Rossi, S.K. Mack, and R.L. Michaels, "Single-point diamond turning and replication of visible and near-infrared diffractive optical elements," *Appl. Opt.* **36**, pp. 4648–4654 (1997).

31. Y. Hashimoto, Y. Takeuchi, T. Kawai, K. Sawada et al., "Manufacture of fly-eye mirror in an extreme-ultraviolet lithography system by means of ultraprecision diamond cutting," in *Emerging Lithographic Technologies VI*, R.L. Engelstad, Ed., *Proc. SPIE* **4688**, pp. 656–663 (2002).

32. J. Ruckman, H. Pollicove, and D. Golini, "Advanced manufacturing generates conformal optics," *Optoelectron. World* (7), pp. S13–S15 (1999).

33. S.M. Shank, M. Skvarla, F.T. Chen, H.G. Craighead et al., "Fabrication of multi-level phase gratings using focused ion beam milling and electron beam lithography," in *Diffractive Optics*, Vol. 11 of OSA Technical Digest Series. Optical Society of America, Washington, DC, pp. 302–304 (1994).

34. P. Kung and L. Song, "Rapid prototyping of multi-level diffractive optical elements," in *Diffractive Optics*, Vol. 11 of OSA Technical Digest Series. Optical Society of America, Washington, DC, pp. 133–136 (1994).

35. M.T. Duignan, "Micromachining of diffractive optics with excimer lasers," in *Diffractive Optics*, Vol. 11 of OSA Technical Digest Series. Optical Society of America, Washington, DC, pp. 129–132 (1994).

36. G.P. Behrmann and M.T. Duignan, "Excimer laser micromachining for rapid fabrication of diffractive optical elements," *Appl. Opt.* **36**, pp. 4666–4674 (1997).

37. X. Wang, J.R. Leger, and R.H. Rediker, "Rapid fabrication of diffractive optical elements by use of image-based excimer laser ablation," *Appl. Opt.* **36**, pp. 4660–4665 (1997).

38. F.P. Shvartsman, "Replication of diffractive optics," in *Diffractive and Miniaturized Optics*, S.H. Lee, Ed., *SPIE Press* **CR49**, pp. 117–137 (1993).

39. M.T. Gale, "Replication," in *Micro-optics: Elements, Systems, and Applications*, H.P. Herzig, Ed., Taylor and Francis, London, pp. 153–177 (1997).

40. J.-L.R. Nogues and R.L. Howell, "Fabrication of pure silica micro-optics by sol-gel processing," in *Miniature and Micro-Optics: Fabrication and System Applications II*, C. Roychoudhuri and W.B. Veldkamp, Eds., *Proc. SPIE* **1751**, pp. 214–224 (1992).

41. Y. Xia and G.M. Whitesides, "Soft lithography," *Angew. Chem. Int. Ed.* **37**, pp. 550–575 (1998).

42. J.L. Wilbur, A. Kumar, E. Kim, and G.M. Whitesides, "Microfabrication by microcontact printing of self-assembled monolayers," *Advan. Mater.* **6**, pp. 600–604 (1994).

43. P.M. St. John and H.G. Craighead, "Microcontact printing and pattern transfer using trichlorosilane on oxide substrates," *Appl. Phys. Lett.* **68**, pp. 1022–1024 (1996).

44. X. Zhao, Y. Xia, and G.M. Whitesides, "Fabrication of three-dimensional micro-structures: Microtransfer molding," *Advan. Mater.* **8**, pp. 837–840 (1996).

45. P. Nussbaum, I. Philipoussis, A. Husser, and H.P. Herzig, "Simple technique for replication of micro-optical elements," *Opt. Eng.* **37**, pp. 1804–1808 (1998).

46. T.J. Suleski, B. Baggett, H. Miller, B. Delaney et al., "Wafer-scale replication of glass micro-optics for optical communications," in *Diffractive Optics and Micro-Optics*, OSA Technical Digest Series. Optical Society of America, Washington, DC, pp. 231–233 (2000).

47. A. Lopez and H.G. Craighead, "Subwavelength surface-relief gratings fabricated by microcontact printing of self-asssembled monolayers," *Appl. Opt.* **40**, pp. 2068–2075 (2001).

Chapter 9

Testing Diffractive Optical Elements

Once a surface relief pattern has been patterned in photoresist or etched into a substrate, how do you know if you have produced what you wanted? There are many different approaches to answering this question, but all of them fall into one of two categories: (1) metrological tests that measure the dimensions and geometry of the surface structures or (2) tests that measure the optical performance of the component. Both types of tests are necessary. Although dimensional measurements will provide information on the fidelity of the fabrication process, they cannot assure you that the element works properly. (Suppose someone goofed and the design is incorrect?) We now consider a number of techniques for both dimensional and optical testing of diffractive optical elements.

9.1 Metrology

Dimensional measurements of diffractive optic structures are used both during and after fabrication of the element. Typical measurements include lateral feature sizes, locations of the transition points across the field of the DOE, grating depths, verticality of grating side walls, rounding of edges, surface roughness, and other geometrical factors. Although the performance of the DOE can usually be inferred from the measured geometry, metrological measurements are normally used to troubleshoot and quantify problems in the fabrication process. We now consider several common methods for the measurement of DOE geometries.

9.1.1 Optical microscopy

Microscopes are perhaps the most commonly used tools for inspection of diffractive optical elements. Although these tools may not provide highly precise measurements, they are relatively inexpensive and easily available. Microscopes are typically used to find gross defects in pattern geometries and for rapid relative measurements of lateral feature dimensions. In many cases, the eyepiece of a microscope will contain a reticle with regularly spaced tick marks. By calibrating the distance between tick marks using a known reference, it is possible to quickly measure feature sizes by estimating the number of tick marks covered by the feature. In some cases, the magnified pattern is imaged by a video camera onto a monitor. A set of electronic lines on the video screen that overlay the image of the structure can be used as "video calipers" that are controlled by the user. The distance between the lines is calibrated using a known reference. This system permits rapid and precise measurements of feature widths. A sample microscope image of a diffractive optical element measured with a video caliper is shown in Fig. 9.1.

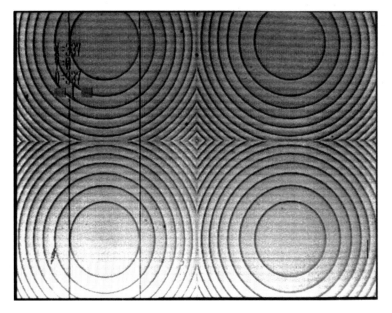

Figure 9.1 Measurement of diffractive lens array using an optical microscope and "video caliper" arrangement.

Other optical inspection methods can also be used, depending on the need. Optical comparators were traditionally used for measurement of spectral lines on film plates. These devices determine the distance a microscope stage traveling on a precision screw has moved by using the pitch of the thread. Image-processing software can also be used for geometrical measurement and analysis.

9.1.2 Mechanical profilometry

After optical microscopy, mechanical profilometry is probably the most common dimensional measurement method for diffractive optical elements. A profilometer is a delicate instrument that moves a sharp stylus over the surface of the DOE and records the movement of the stylus. The stylus tip can be a ball at the tip of a cone or a bar (like the edge of a prism), depending on the geometry to be measured (bar geometries are useful for measuring refractive microlens shapes because they ensure that the stylus traces over the apex of the lens). These tools can measure the depth of the structure and also provide information on the smoothness of the surface. The scan range of this method is adjustable from micrometers to millimeters. Care must be taken not to damage the surface being scanned, particularly if it is a relatively soft material like photoresist. The force exerted by the stylus tip must not damage the piece under test. Similarly, too light a force will cause the stylus tip to "skip" across the surface and decrease the accuracy of the measurement. Profilometer scans of three diffractive elements and a refractive microlens are shown in Fig. 9.2.

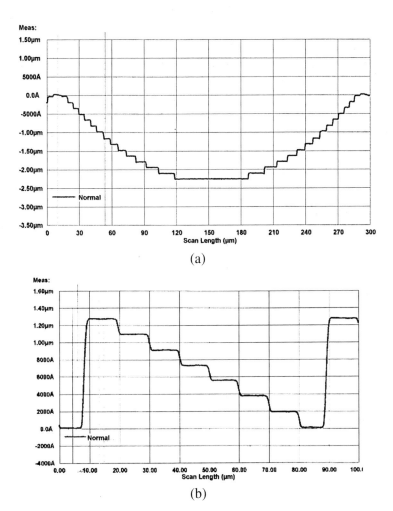

Figure 9.2 Mechanical profilometry scans of diffractive and refractive micro-optics. (a) Sixteen-level lens. (b) Eight-level blazed grating.

When "large" features are measured, profilometers give very accurate information on feature depths (depth accuracies of about 10 Å are achievable). Although the measurements of feature widths and transition locations are less accurate, they give a reasonable representation of the surface shape. Most profilometers generate a one-dimensional scan of the surface, although newer machines can perform a

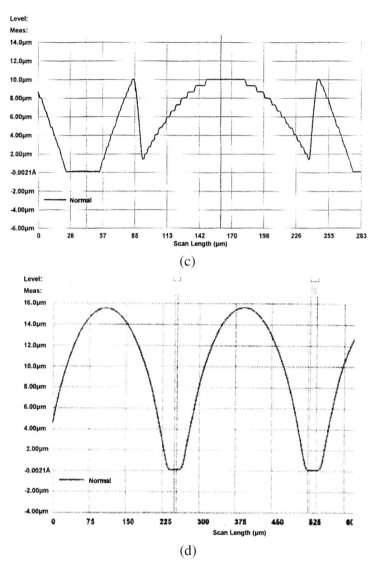

(c)

(d)

Figure 9.2 (Continued.) (c) Harmonic diffractive lens. (d) Refractive microlens. (All figures courtesy of Digital Optics Corp., Charlotte, NC.)

series of 1D scans to build up a 2D contour of the element. A key issue with profilometer measurements is the relationship between the accuracy of the measured surface profile and the size and shape of the stylus tip. In reality, the measured profilometer trace is a convolution of the actual profile and the stylus tip geometry. The trace profile generated by a mechanical profilometer shows the path traversed by the center of curvature of the stylus tip, *not* the actual surface profile. This distinction must be kept in mind when interpreting profilometry traces. When the feature being measured is comparable in size to the tip size, as shown in Fig. 9.3(a), the tip will not reach the bottom of the "trench," resulting in an erroneous depth measurement. The error is dependent on the type of detail that is being probed. Similarly, the shank angle of the stylus tip can ride over the sharp corners of a feature, resulting in a profilometry trace that looks significantly different than the actual surface [Fig. 9.3(b)]. Typical tip radii range from 2 to 12.5 μm, and the shank angle can range up to 120 deg or more.

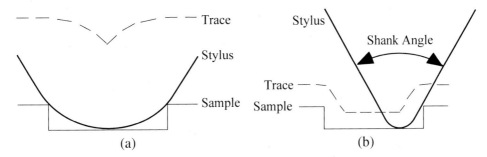

Figure 9.3 Mechanical profilometry of diffractive optics with small features.

9.1.3 Atomic force microscopy

Atomic force microscopy is another method that is commonly used to measure the geometries of diffractive optical elements. Similar to mechanical profilometry, AFM uses a very fine-tipped stylus to measure the features. Unlike mechanical profilometry, the AFM stylus never comes directly in contact with the sample (except by accident!). Instead, the AFM measures small forces between the atoms in the sample and the atoms in the stylus tip. By measuring these forces with a small tip on a cantilevered arm with a feedback mechanism, it is possible to measure the geometry of the surface shape with great accuracy. AFM measurements have been used in the past to image individual atoms and molecules. Like mechanical profilometry, the lateral resolution of the measurement can be limited by the stylus shape "riding" over the edges of the features under measurement, although stylus tips for AFMs are much smaller (tip sizes with radii <10 nm, for example, with small shank angles). A wide range of tip shapes are used with AFMs to achieve different effects. For example, long, narrow tips with sharp, flared edges can be used to more accurately measure the side-wall shapes of submicrometer features. The higher resolution of an AFM also limits the measurement area; practical scan ranges are from <1 μm to tens of microns. AFMs can be used to scan in one dimension or in two dimensions, as shown in Fig. 9.4. Additional information on the operating principles of AFMs can be found in a number of texts.[1,2]

9.1.4 Scanning electron microscopy

Scanning electron microscopy can also be used to determine the geometries of diffractive optical elements. One of the biggest advantages of SEM measurement is its incredibly high resolution. Magnifications of 60,000 times or more can be readily achieved. Furthermore, the resolution can be changed relatively easily once

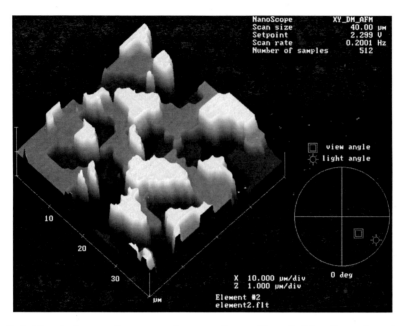

Figure 9.4 Atomic force microscope image of a multilevel diffractive microstructure that creates a uniform array of light spots. (Courtesy of Digital Optics Corp., Charlotte, NC.)

the sample area is in focus by simply twisting a knob. SEM images of several diffractive optical elements are shown in Fig. 9.5. The principles that govern an SEM are similar to those of conventional microscopes in that high-quality "optics" are used to image the sample. In this case, however, the wave nature of matter is the operating principle, rather than the wave nature of light. The "optics" in the SEM that redirect and shape the electron beam used for imaging the sample are electrical. (The wavelength of an electron is approximately 0.4 Å when the accelerating voltage of the electron is 1000 V. The wavelength decreases as the inverse square of the accelerating voltage.[1])

Although the SEM measurements are noncontact, they can sometimes require the destruction of the sample being measured. In many cases, it is necessary to overcoat the sample with a conductive layer (usually a few hundred angstroms of gold or platinum) to prevent a buildup of electrical charge from the electron beam that would cause the image to degrade. A "field emission" SEM makes it possible to measure to samples without the need for the conductive layer by imaging with low voltage beams.

SEM measurements can be particularly powerful when they are coupled with the use of a focused ion beam described in Sec. 8.2.3. The FIB can be used to cut a cross section in the sample being measured. Then, by using the SEM, an extremely precise measurement of the grating cross section can be obtained [see Fig. 9.5(a)]. This technique is particularly good for accurate measurements of critical dimensions and side-wall angles of diffractive features. One caveat: it is necessary to take the viewing angle into account when determining the absolute dimensions of the

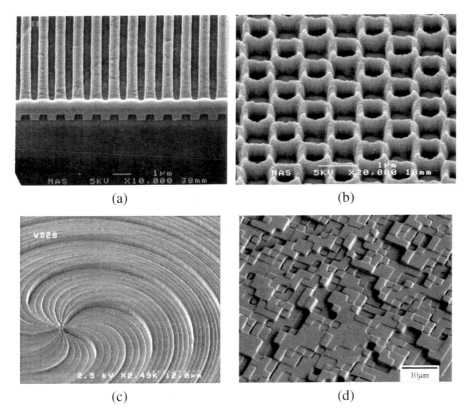

Figure 9.5 SEM images of a variety of diffractive optical elements. (a) Binary grating with 0.5-μm features. (b) Two-dimensional binary grating with sub-0.5-μm features. (c) Eight-phase level diffractive beam shaper. (d) Eight-phase level diffractive spot array generator. (All figures courtesy of Digital Optics Corp., Charlotte, NC.)

features being measured. In order to see 3D topology, the sample must be viewed at non-normal incidence—the apparent feature dimension must be scaled appropriately with the viewing angle against a known dimensional reference to get an accurate measurement.

9.1.5 Phase-shifting interferometry

Phase-shifting interferometry is an extremely powerful, noncontact technique for the measurement of optical wavefronts and shapes.[3] Dating back to the late 1960s and early 1970s, this technique has been used to measure the surface topographies of diffractive optical elements in recent years. Using this method, a time-varying phase shift is introduced between a reference wavefront and the test wavefront. The wavefront phase is encoded in the intensity patterns of the recorded interferograms. The phase (and then the height) of the test element is recovered through a series of point-by-point calculations. This technique has a vertical resolution measured in angstroms and a lateral resolution on the order of 1-μm (the lateral resolution depends on the wavelength of the light source and imaging optics used in the inter-

ferometry system). Similar in concept to the tip of a mechanical surface profilometer, the finite size of the focused laser beam limits the ability to accurately measure
small features. In addition, the measurement and definition of the vertical features
(side walls) can be limited if the numerical aperture of the microscope objective
is not sufficiently high. A sample measurement of a diffractive lens structure performed using phase-shifting interferometry is shown in Fig. 9.6.

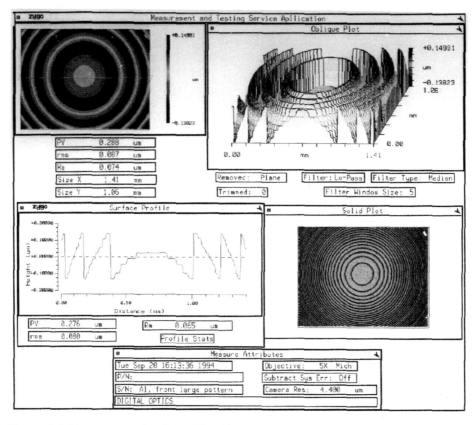

Figure 9.6 Measurements of a diffractive lens with phase-shifting interferometry.
(Courtesy of Digital Optics Corp., Charlotte, NC.)

9.2 Testing Optical Performance

The ultimate test of a fabrication process is the performance of the optics that are
made. There are several ways to measure the performance of diffractive optical
elements. For example, the irradiance patterns generated by DOEs can be measured directly. Diffraction efficiencies can be measured either directly (using a test
laser at the design wavelength of the DOE), or indirectly (using a test laser at a
different wavelength and inferring the performance of the component at the design
wavelength using mathematical models).

The optical testing methods presented here are representative of methods used
to measure the output of diffractive optical elements, but are not a comprehensive

test set. Other measurement methods may be required, depending on the intended function of the DOE. For example, subwavelength DOEs may require measurement of polarization states of the transmitted light. The proper test setup for a given measurement problem must be determined on a case-by-case basis.

9.2.1 Scatterometer

Although dimensional measurements may provide information on the construction details of a DOE, it is only by testing the performance of the element that the actual performance can be determined. For many elements, this can be done using a scatterometer, a device that measures the amount of light directed to the appropriate place—and also the light that goes where it isn't supposed to go! A schematic of the device is shown in Fig. 9.7.

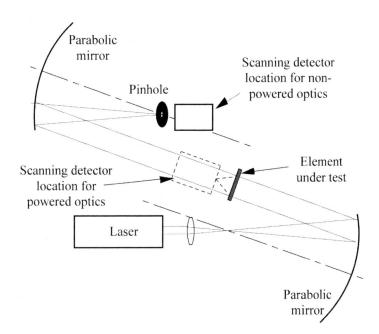

Figure 9.7 Scatterometer setup for measuring the near-field and far-field irradiance patterns from diffractive optical elements.

In this setup, a collimated input is provided to the DOE under test using off-axis illumination from a set of parabolic mirrors. This approach measures the irradiance patterns from DOEs by scanning a detector and pinhole in either the near field or the far field of the test component. The scanning and data acquisition are computer controlled (using LabView software, for example). The precision of the measurement depends on the size of the pinhole and the step size of the motorized stages. Also, the scanning process can be slow, and it can be difficult to properly align the scan axis. Still, this approach can provide detailed measurements of the DOE output (see Fig. 9.8).

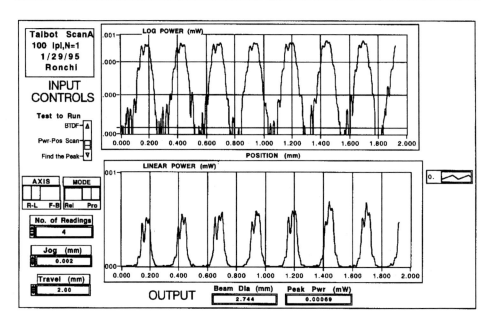

Figure 9.8 Near-field diffraction pattern measured from a periodic grating using a scatterometer and LabView software package with a 2-μm pinhole.

9.2.2 Charge-coupled devices

Charge-coupled device (CCD) sensor arrays provide an easy way to rapidly measure the outputs of diffractive optical elements by providing one- or two-dimensional images of DOE irradiance distributions in real time. The pixels in the array generate an electrical pulse in proportion to the intensity of the light hitting each pixel. This electrical signal is then converted ("digitized") into an integer number that typically ranges from zero (no light) to the upper end of the system's dynamic range for very intense light. Setups with 256 to 65,536 distinct levels of gray-scale resolution (8-bit to 16-bit range) are common. The numbers representing the light intensity illuminating each of the pixels in the array are then used to reconstruct an image of the light pattern hitting the CCD array. This digitized representation of the light pattern lends itself to computer analysis.

To measure the output of a diffractive optical element, the sensor array is placed directly in the path or focal plane of the device under test. Although CCD arrays can provide many benefits for rapid measurements of DOE performance, there are several issues that must be addressed. The typical size of a CCD array is only a few millimeters on a side, which limits the area that can be measured at one time. Also, the measurement precision is limited by the pixel size, although it is possible to use additional imaging optics to increase the effective resolution of the measurement setup. The dynamic range and wavelength sensitivity of the CCD array and digitizing board must also be taken into consideration. Some sample measurements of DOE outputs using CCD arrays are shown in Fig. 9.9. The images produced by the CCD are readily analyzed using image-processing software or

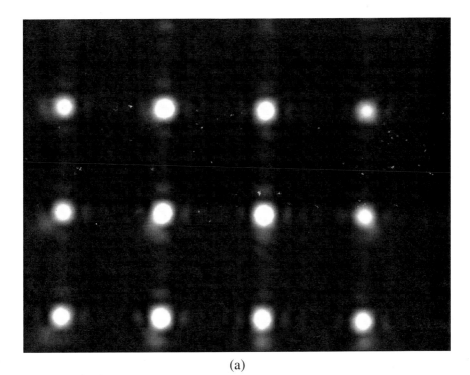

(a)

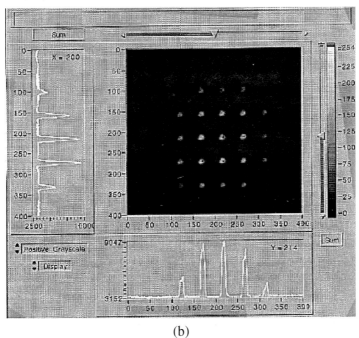

(b)

Figure 9.9 CCD images of the outputs of different types of diffractive optical elements. (a) Focal plane of a microlens array on 200-μm spacing. (b) Far-field output of a 3 × 3 diffractive spot array generator measured with CCD array and LabView software. (Figure (a) courtesy of Digital Optics Corp., Charlotte, NC; Fig. (b) courtesy of Scott Thornburg, Georgia Institute of Technology.)

other techniques to determine the performance properties of the DOE, particularly the uniformity of the output pattern.

9.2.3 Rotating slit scanners

Several methods for measuring the output of diffractive optical elements in the near and far fields have been presented. Another method makes use of rotating slit scanners (RSS) that are normally used to characterize laser beam profiles. However, these devices can also be used to measure outputs from diffractive optical elements with great precision. Rotating slit scanners consist of a long, narrow slit (typically several millimeters long and several micrometers wide) on a rotating drum. It is possible to obtain slit widths as small as 0.5 to 1.0 μm. A photodetector inside the drum is illuminated as the clear slit passes in front of it between 10 and 20 times per second. As the slit rotates, the irradiance distribution from the DOE is sampled across an area several millimeters in extent. The resulting output from the detector is an irradiance profile that is several millimeters across, sampled at micrometer intervals. A sample setup for measurement of diffraction patterns using an RSS is shown in Fig. 9.10.

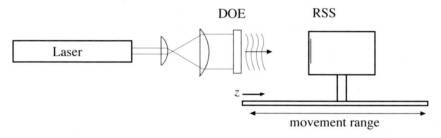

Figure 9.10 Experimental geometry for measurement of DOE output using a rotating slit scanner.

The use of rotating slit scanners has several advantages over CCD arrays for measuring the output of diffractive optical elements. The RSSs can have higher measurement resolution than CCD arrays. They can provide irradiance profile data in a format that is easily imported into mathematical software for analysis and visualization. However, rotating slit scanners are more limited than CCDs when it comes to measuring two-dimensional patterns. The resolution of the RSS is limited for 2D scans by the length of the slit. As the slit rotates, the detector integrates all the energy passing through the entire slit at each sampling point. This integration does not present a problem when measuring 1D or circular irradiance distributions, but it can give misleading results when measuring discrete 2D irradiance patterns. Also, the RSS slit traverses a curved path. As a result, the lateral irradiance distribution is sampled over a small longitudinal range as the slit rotates, whereas a CCD measures the output in a flat plane. Still, high-precision measurements of DOE outputs are readily obtainable. A sample RSS measurement of a near-field diffraction pattern from a DOE is shown in Fig. 9.11.

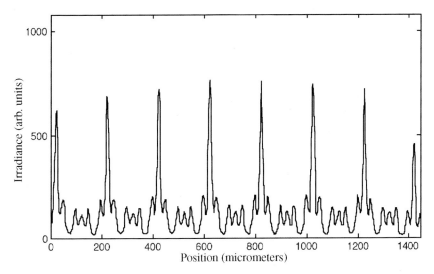

Figure 9.11 Near-field irradiance pattern measured 15.8 mm from a 1 × 7 Dammann grating with a 200-μm period tested at 632.8 nm.

9.2.4 Array testing

As an example of a test geometry for a diffractive spot array generator, consider the example shown in Fig. 9.12. In this setup, a detector with a pinhole aperture is mounted on a translation stage under computer control. By scanning the detector to the center of each of the spots, it is possible to measure the power contained in each of the spots (diffraction orders) in the output array. A sample measurement using this system is shown in Fig. 9.13(b). In a similar manner, the detector can be mounted on a fixed post and a wafer containing many individual diffractive elements can be mounted on the translation stage. By translating the wafer so that each element is placed in front of the laser in turn, the performance of each of the

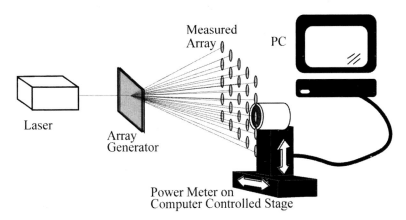

Figure 9.12 Schematic of apparatus used for diffractive array measurements. (Courtesy of Digital Optics Corp., Charlotte, NC.)

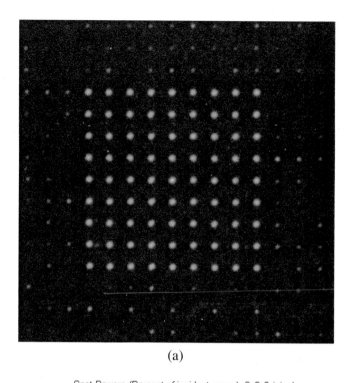

(a)

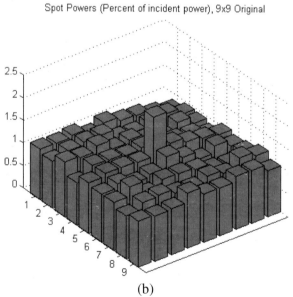

(b)

Figure 9.13 (a) A CCD image of the far-field output of a 9 × 9 spot array generator. (b) Output of a 9 × 9 spot array generator designed for 650-nm light but tested at 670 nm using the setup shown in Fig. 9.12. (Both figures courtesy of Digital Optics Corp., Charlotte, NC.)

parts on the wafer can be tested. Obviously, this is a time-consuming process, but it provides a comprehensive testing of an element or an array of elements.

9.3 Effects of Fabrication Errors on DOE Performance

The relationship between the shape of a diffractive optical element and its performance is generally well known; this relationship is the basis of the design and fabrication procedures described in detail in earlier chapters. However, only rarely are fabricated DOEs identical to the ideal design. In reality, there is always some degree of deviation from the ideal shape in the fabricated geometry. It is enlightening to consider the effects of these fabrication errors on the performance of diffractive optical elements. Using the same types of analysis discussed in earlier chapters, it is possible to quantitatively assess the effects of fabrication errors on the performance of diffractive optical elements. Both theoretical and experimental studies of these effects have been performed.[4–12] There are many types of errors that can occur during processing. For example:

- Etch-depth errors
- Feature-size errors
- Rounding of sharp corners
- Transition-point errors
- Grating-period deviations

- Nonvertical side-wall angles
- Mask-to-mask alignment errors
- Increased surface roughness
- Shape errors arising from analog processing

Most of these errors are systematic in that they occur in the same manner throughout the structure. These systematic errors affect the distribution of light into the various orders of grating structures. Other errors, such as surface roughness, occur more randomly and do not affect the relative intensities of the orders. Instead, all of the orders are robbed of some intensity and the lost light is redistributed into the background between orders. Some of these error types and their effects on grating performance are illustrated in Fig. 9.14.

A discussion of the effects of variation in phase depth and grating duty cycle for a simple binary phase grating (of the type shown in Fig. 6.1) was presented in Sec. 5.1.2. We briefly consider the effects of small fabrication errors on this grating. As a more complex example of the effects of fabrication errors on DOE performance, we also consider the scalar diffraction efficiency of a binary phase, 1×5 Dammann grating designed to give equal energy in the central five diffraction orders.

When the simple grating from Fig. 6.1 has a phase depth of π and a duty cycle of 50%, none of the transmitted energy goes into the zeroth order, and 40.5% of the transmitted light goes into the $+1$ and -1 orders (each). It is interesting to explore what happens to the grating performance when either the phase depth is changed, or the grating duty cycle is changed. Plots of these effects are illustrated in Fig. 5.3 and Fig. 5.4. For this simple grating, depth errors have a relatively minor effect on the grating efficiency and uniformity. For the 50% duty cycle grating, etch depth

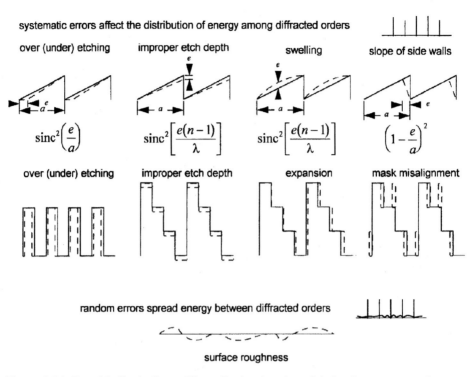

Figure 9.14 Graphic illustration of the effects of various fabrication errors on the performance of diffractive optical elements.[4]

errors of ±10% decrease the ±1 first-order diffraction efficiency from 40.5% to about 39.5% while raising the power in the zeroth order from 0 to about 2%. For this grating, variations in the duty cycle also cause relatively minor effects.

In comparison, consider the behavior of a binary phase 1 × 5 Dammann grating, as shown in Fig. 9.15(a). When this grating is fabricated perfectly, 77.4% of the total transmitted light is directed into the central five diffraction orders, with 15.48% of the light in each of these orders, as shown in Fig. 9.15(b). A 10% depth error in the grating results in 17.6% of the light in the zeroth order and 15.1% of the light in each of the other four central orders, as shown in Fig. 9.16(a). Changing the location of one of the grating transitions has an even more profound effect on the grating output, as shown in Fig. 9.16(b).

This type of behavior is typical for beam fanouts. For binary phase structures for which scalar diffraction theory is applicable, etch depth errors result in an increase in the power in the zeroth order, with a corresponding, uniform decrease in power in the other diffraction orders. In comparison, changes in the locations of phase transitions cause the light distributions between the diffraction orders to vary nonuniformly.[5] For multiphase level beam fanouts, etch depth errors result in additional nonuniformity across all diffraction orders. The light distribution in all of the diffraction orders for multilevel fanouts is also affected by transition location and mask misalignment errors.[9]

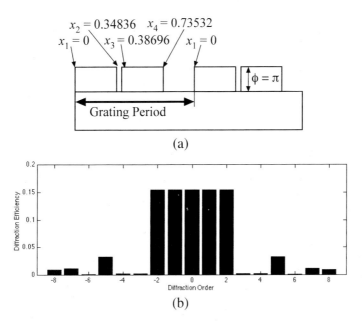

(a)

(b)

Figure 9.15 (a) Two periods of a Dammann grating designed to direct equal amounts of energy into the central five diffraction orders. The locations of grating transitions are given with a normalized grating period. (b) The output of the grating in (a).

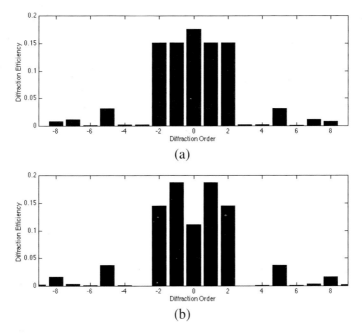

(a)

(b)

Figure 9.16 (a) The output of the grating shown in Fig. 9.15(a) when the etch depth is increased by 10%. (b) The output of the grating shown in Fig. 9.15(a) when the transition x_4 is changed from 0.73532 to 0.70532.

Different types of DOEs are affected differently by the various types of fabrication errors. We have given several examples for periodic grating structures. Aperiodic structures (such as diffractive lenses) are also affected by fabrication errors, although in somewhat different ways. Although fabrication errors on these structures also redistribute power between the various diffraction orders of the lens (see discussion in Sec. 4.3), the net effect is a decrease in the amount of power in the focal spot and a decrease in the Strehl ratio of the lens.[7]

It should also be noted that the choice of fabrication method can have a significant effect on the types of fabrication errors that may occur. For example, grayscale lithography and direct writing tend to have nonvertical side walls at sharp phase transitions, and variations in exposure dosage can cause shape errors for both, but these techniques do not suffer from mask-to-mask alignment errors. Conversely, binary optics fabrication techniques tend to have very sharp side walls and are much less susceptible to variations in exposure dosage, but one must be aware of the effects of mask misalignment errors. In this regard, the type, effect, and likelihood of fabrication errors must be weighed against cost and performance requirements when choosing which of the fabrication techniques discussed in Chapters 7 and 8 will be used for a given task.

References

1. J.R. Sheats and B.W. Smith, *Microlithography Science and Technology*. Marcel Dekker, New York (1998).
2. L.J. Lauchlan, D. Nyyssonen, and N. Sullivan, "Metrology methods in photolithography," in *Handbook of Microlithography, Micromachining, and Microfabrication, Vol. 1: Microlithography*, P. Rai-Choudhury, Ed., SPIE Press, Bellingham, WA, pp. 475–595 (1997).
3. J.E. Greivenkamp and J.H. Bruning, "Phase shifting interferometry," in *Optical Shop Testing*, D. Malcara, Ed. Wiley, New York, pp. 501–598 (1992).
4. T. Fujita, H. Nishihara, and J. Koyama, "Blazed gratings and Fresnel lenses fabricated by electron-beam lithography," *Opt. Lett.* **7**, pp. 578–580 (1982).
5. J. Jahns, M.M. Downs, M.E. Prise, N. Streibl et al., "Dammann gratings for laser beam shaping," *Opt. Eng.* **28**, pp. 1267–1275 (1989).
6. J.A. Cox, T. Werner, J. Lee, S. Nelson et al., "Diffraction efficiency of binary optical elements," in *Computer and Optically Formed Holographic Optics*, I. Cindrich and S.H. Lee, Eds., *Proc. SPIE* **1211**, pp. 116–124 (1990).
7. M.W. Farn and J.W. Goodman, "Effect of VLSI fabrication errors on kinoform efficiency," in *Computer and Optically Formed Holographic Optics*, I. Cindrich and S.H. Lee, Eds., *Proc. SPIE* **1211**, pp. 125–136 (1990).
8. M.B. Stern, M. Holz, S.S. Medeiros, and R.E. Knowlden, "Fabricating binary optics: process variables critical to optical efficiency," *J. Vac. Sci. Technol. B* **9**, pp. 3117–3121 (1991).
9. J.M. Miller, M.R. Taghizadeh, J. Turunen, and N. Ross, "Multilevel-grating array generators: fabrication error analysis and experiments," *Appl. Opt.* **32**, pp. 2519–2525 (1993).

10. D.A. Pommet, E.B. Grann, and M.G. Moharam, "Effects of fabrication errors on the diffraction characteristics of binary dielectric gratings," *Appl. Opt.* **34**, pp. 2430–2435 (1995).
11. T.J. Suleski and D.C. O'Shea, "Gray-scale masks for diffractive-optics fabrication: I. Commercial slide imagers," *Appl. Opt.* **34**, pp. 7507–7517 (1995).
12. D.C. O'Shea, "Reduction of the zero-order intensity in binary Dammann gratings," *Appl. Opt.* **34**, pp. 6533–6537 (1995).

Chapter 10

Application of Diffractive Optics to Lens Design

10.1 Introduction

In Chapter 4 the diffractive lens, including a simple hybrid achromat, was discussed in some detail. In this chapter, we extend our discussion of lens design by introducing the wavefront aberration polynomial and the Seidel coefficients. Although it is not our intention to sneak a lens design text into this book, we do have to provide sufficient information on the expression of aberrations and their correction so that you can gain an appreciation of the power of diffractive surfaces in lens design and the additional degrees of freedom that they provide. The limitations of these approaches are also discussed. The chapter begins with an introduction to the aberration polynomial and its expression for a thin lens. Then pure diffractive designs, including superzone versions, are discussed. This is followed by a number of examples of hybrid lenses. Diffractive surfaces can also be used in the design of systems with reduced thermal sensitivity. The chapter concludes with a short discussion of applications that use microlenses.

10.1.1 The aberrations of a diffractive lens

In Chapter 4 the third-order aberrations (spherical, coma, astigmatism, field curvature, and distortion) were introduced and graphic examples were given for each one. These aberrations[1] arise when a wavefront propagated through the optical system is not a perfect replica of the initial wavefront. This can be described as a function of a normalized object height h and the normalized pupil coordinate, ρ and θ, as shown in Fig. 10.1. A wavefront aberration [Eq. (10.1)] can be expanded to any number of even orders of the powers of h and ρ, but the most important are those of the fourth power of the product of h and ρ:

$$
W(h, \rho, \cos\theta) = \frac{1}{8}S_I\rho^4 + \frac{1}{2}S_{II}h\rho^3\cos\theta + \frac{1}{2}S_{III}h^2\rho^2\cos^2\theta
$$
$$
+ \frac{1}{4}\left(S_{III} + S_{IV}\right)S_{II}h^2\rho^2 + \frac{1}{2}S_Vh^3\rho\cos\theta
$$

(10.1)

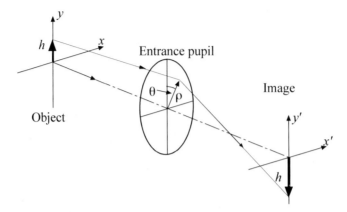

Figure 10.1 Coordinate geometry for describing the aberrations of an optical system.

The coefficients of the polynomial, S_{I-V}, are the Seidel coefficients: spherical aberration, coma, astigmatism, field curvature, and distortion, respectively. These are referred to as third-order aberrations because the ray error is derivative of Eq. (10.1) with respect to h and ρ, which yields a third-order expression.

Because we will be applying these concepts to diffractive optics, a first simplification is to determine the expression for a thin lens, which has a refractive index n and two curvatures, c_1 and c_2, but no thickness. Two dimensionless parameters that can be defined are the bending parameter B in term of the curvatures and index, and the conjugate parameter T in terms of the magnification m at which the lens will be used ($m = 0$, if the object is at infinity):

$$B = \frac{c_1 + c_2}{(n-1)(c_1 - c_2)} = \frac{c_1 + c_2}{\phi} \quad \text{and} \quad T = \frac{m+1}{m-1}. \qquad (10.2)$$

The Seidel coefficients for a thin lens of power $\phi = (n-1)(c_1 - c_2)$, and an axial ray height on the lens of y are given by

spherical aberration

$$S_I = \frac{y^4 \phi^3}{4}\left[\left(\frac{n}{n-1}\right)^2 + \frac{n+2}{n}B^2 + \frac{4(n+1)}{n}BT\right.$$
$$\left. + \frac{3n+2}{n}T^2\right], \qquad [10.3(a)]$$

coma

$$S_{II} = \frac{-y^2\phi^2 H}{2}\left[\frac{n+1}{n}B + \frac{2n+1}{n}T\right], \qquad [10.3(b)]$$

astigmatism

$$S_{III} = H^2\phi, \qquad [10.3(c)]$$

field curvature

$$S_{IV} = \frac{H^2\phi}{n}, \quad \text{and} \qquad [10.3(d)]$$

distortion

$$S_V = 0, \qquad [10.3(e)]$$

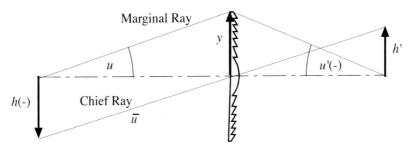

Figure 10.2 Propagation of a marginal ray and a chief ray through a single diffractive lens. The optical invariant H for this simple lens is given by $-y\bar{u}$. The negative signs in parentheses indicate that the corresponding quantities are negative.

where the quantity H is the optical invariant, which for the thin-lens case is equal to $-y$ times the chief ray angle $\bar{u}$, the angle the fractional object height h makes with the thin lens (Fig. 10.2).

10.1.2 Adapting optical design for diffractive elements: the Sweatt model

Although a diffractive lens certainly is thin, it really isn't a lens in the conventional sense. As noted in Sec. 4.2.1, a diffractive lens is actually a variable frequency grating. A number of approaches to modeling a diffractive structure have been constructed, but one of the simplest is a model devised by William Sweatt.[2] In this model, the diffractive optical element or a holographic optical element (HOE) is treated as a material with a high refractive index on a substrate. This simple strategy permits a designer to model a lens within a standard optical design program that has no special provision for treating DOEs.

For a thin wedge of material with a high refractive index (Fig. 10.3), it can be shown that

$$(n-1)dt/dr = \sin\theta_d - \sin\theta_i. \tag{10.4}$$

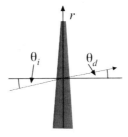

Figure 10.3 The geometry for the derivation of the Sweatt model.

And for a diffraction grating (Sec. 4.2.1) whose period varies across the surface as $\Lambda(r)$,

$$m\lambda/\Lambda(r) = \sin\theta_d - \sin\theta_i, \quad (10.5)$$

where λ is the wavelength and m is the diffraction order. Equating

$$\sin\theta_d - \sin\theta_I = m\lambda/\Lambda(r) = (n-1)dt/dr. \quad (10.6)$$

For large refractive indices ($n \gg 1$), we can replace $n - 1$ by n:

$$\sin\theta_d - \sin\theta_i = m\lambda/\Lambda(r) = ndt/dr. \quad (10.7)$$

The change in the angle of the rays passing through the diffractive structure can be expressed either in terms of the of the grating parameters $[m\lambda/\Lambda(r)]$ or in terms of the optical path difference along the surface of an equivalent lens at that point $[ndt/dr]$. Note that neither the center nor the right-hand side term of Eq. (10.7) is a function of the angles of incidence. They are only a function of the location r on the surface where the ray enters.

In the case of a lens, its variation in thickness with radius using the sag equation is $t(r) = r^2/2R$. The derivative of thickness with respect to the radius is then $dt/dr = 2r/2R = r/R$. Substituting this into the right-hand side of Eq. (10.7), $m\lambda/\Lambda(r) = (n-1)dt/dr$, the periodicity of the grating for a first-order ($m = 1$) diffraction is $\Lambda(r) = \lambda/(n-1)dt/dr = \lambda R/(n-1)r$. But $R/(n-1)$ is the focal length of a plano-convex lens. Therefore,

$$\Lambda(r) = f\lambda/r. \quad (4.15)$$

This same expression was derived using Fresnel zones in Sec. 4.2.1.

Consider the simple example of a diffractive lens modeled as a high-index, 100-mm EFL plano-convex lens. The focal length of the lens is given by Eq. (4.4): $f = R/(n-1)$. If the refractive index is set at 10,000, then the radius of curvature of the curved surface is $R = 10^4 \times 100$ mm $= 10^6$ mm $= 10^3$ m. The thickness of a 20-mm diameter lens, using the sag equation is

$$t(= r^2/2R) = (10 \text{ mm})^2/2 \times 10^6 \text{ mm} = 0.05 \text{ } \mu\text{m} = 50 \text{ nm}.$$

With the Sweatt model, the lens is a thin structure with the correct focal length and phase. The aberrations of a DOE also can be determined using this approach. In this example we have used a spherical surface, but it is possible to include aspheric terms in the surface description.

10.2 Diffractive Lenses

If we treat a thin lens as a diffractive structure, we can determine the Seidel coefficients given in Eq. (10.3) by letting n become large. However, the bending parameter B of the lens [Eq. (10.2)] also would become infinite if we use the first of the definitions in Eq. (10.2). However, if the second is used [$B = (c_1 + c_2)/\phi$] and we let ϕ equal the power for the diffractive lens that was given in Chapter 4, the curvatures of the two surfaces, c_1 and c_2, become equal to the curvature of the substrate c_s to which the diffractive surface is applied. This gives the bending parameter B a value of $2c_s/\phi$. In the limit of large n, the Seidel coefficients for the diffractive lens are then reduced to

spherical aberration $\qquad S_I = \dfrac{y^4 \phi^3}{4} \left[1 + B^2 + 4BT + 3T^2\right] - 8\lambda G p y^4$, $\qquad$ [10.8(a)]

where an additional term in the form of a fourth-order coefficient G has been added so that a higher-order term in the phase function can be used to cancel the first term in this expression. The order of the diffraction is p.

Coma becomes $\qquad\qquad\qquad S_{II} = \dfrac{-y^4 \phi^2 H}{2} (B + 2T) \qquad\qquad$ [10.8(b)]

and the astigmatism is given by $\qquad S_{III} = H^2 \phi.$ $\qquad\qquad\qquad$ [10.8(c)]

The field curvature in all cases is zero for a diffractive lens since $S_{IV} \to 0$, as n becomes large and the distortion remains zero.

$\qquad$ The simplest method of correcting this lens with an object at infinity ($T = -1$ since $m = 0$) would be to set the quantity in brackets in Eq. [10.8(a)] equal to zero and not add any fourth-order correction ($G = 0$). In that case, the bracket is equal to $[1 + B^2 - 4B + 3]$ and the spherical aberration is equal to zero if $B = 2$. The coma is also zero, since the bracket in Eq. [10.8(b)] is zero, with $B = 2$ and $T = -1$. In this case, $B = 2c_s/\phi = 2$, requiring that the substrate curvature be equal to the power of the lens. That is, the surface would be concentric with the focal point. With all these simplifications, however, the astigmatism would not be corrected. So while the diffractive profile carries all the optical power, it must be etched onto a curved substrate to reduce S_I and S_{II}.

10.2.1 The $f - \theta$ lens

The ideal performance for a laser scanner is a scanned beam whose focus moves across a flat field, the scanned surface. The system should have the same performance off-axis as it has on-axis. Not only must coma and field curvature be zero, but astigmatism must also be eliminated. This can be done by using an additional degree of freedom of this system, the location of the aperture stop of the system. As noted in Sec. 4.1.1, the aperture stop is the optical element or aperture that limits the amount of light that can enter the optical system. For a simple lens, the edge of the lens itself limits the bundle of light that it can image. But it need not be so. When a separate limiting aperture, called the *stop*, is moved away from the lens, the aberrations change in a predictable manner that is governed by a set of relations called the *stop shift equations*. By moving the stop and changing the phase of the diffractive lens to include the fourth-order phase term, Buralli and Morris[3] show that coma and astigmatism can be eliminated, but that spherical aberration remains and distortion is introduced.

$$S_I = \frac{y^4 \phi^3}{4}(B + 2T)^2,$$ [10.9(a)]

$$S_{III} = S_{IV} = S_V = 0,$$ [10.9(b)]

$$S_V = \frac{2H^2}{y^2(B + 2T)}.$$ [10.9(c)]

In some applications, the introduction of distortion is intentional. For a laser scanner, the beam that is focused onto an image plane by a distortion-free lens will direct the central or chief ray with constant magnification independently of the entering angle. The result is that the laser spot is located at point on the image plane equal to $f \tan\theta$. But as the beam is moved across the scan lens by a rotating scanner, usually a polygon scanner, the angle of the beam with respect to the optic axis θ changes linearly with time. If the beam is to be scanned at a constant rate across the image plane, the scan distance y must also be proportional to the input beam angle. That is, $y = f\theta$. Therefore, a scan lens has to have a built-in distortion to provide the correct beam displacement.

The distortion is the difference between $f\theta$ and $f \tan\theta$. The scan angle θ can be related to the chief ray angle $\bar{u}$ as the product of the $\bar{u}$ times the fractional object height h. To the third order, this amounts to

$$\varepsilon_y = f\theta - f\tan\theta = -\frac{1}{3}f\bar{u}^3 h^3 + \cdots.$$ (10.9)

The distortion error is the derivative of the wavefront aberration polynomial with respect to r. This is equal to

$$\varepsilon_y = \frac{1}{n'u'}\frac{\partial W}{\partial \rho} = \frac{1}{n'u'}\frac{1}{2}S_V h^3,$$ (10.10)

where u' is the final slope angle of the axial ray and n' is the refractive index in image space; i.e., $n = 1$. It can be shown that for an object at infinity, the slope angle is the negative of the axial ray height divided by the focal length: $u' = -y/f$. Substituting for u' in Eq. (10.11), the distortion error may be written as

$$\varepsilon_y = -\frac{f}{2y}S_V h^3.$$ (10.11)

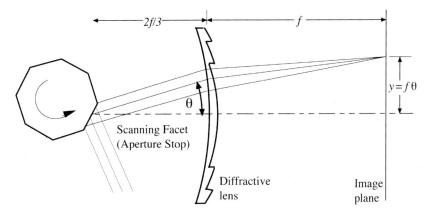

Figure 10.4 The scanning geometry for a laser beam scanner. The radius of curvature of the substrate on which the diffractive lens is located is twice the focal length of the lens, and the scanning facet, the aperture stop of the system, is two-thirds of the focal length from the lens.

Equating these two expressions [Eqs. (10.10) and (10.12)] for distortion determines the Seidel coefficient for distortion S_V required to produce a linear scan. This amounts to $S_V = (2/3)y\bar{u}^3$. When this is equated to the Seidel coefficient for a thin lens with a shifted stop (Eq. [10.9(c)]) and the object is at infinity ($T = -1$), one can solve to find that $B = -1$. Since $B = -2c_s/\phi$, the diffractive element should be placed on a surface whose curvature is $c_s = -\phi/2$ or $R_s = -2f$. From the stop shift equations, the location of the scanner can be determined to be in front of the diffractive lens by two-thirds of its focal length. The complete geometry is shown in Fig. 10.4. Such a scanner using a 150-mm focal length at the He:Ne laser wavelength (633 nm) and a maximum scan angle of ± 20 deg will deliver nearly diffraction-limited performance with a 16-μm spot over an 8.5-inch scan length with a resolution of approximately 1600 dots/inch.

Because it is easier to fabricate a diffractive structure on a flat surface, this design can be reexamined with this simplification present. This change sets $B = 0$ and reintroduces coma, although the $f\theta$ condition is still satisfied. The performance of this simpler design is reduced to about 700 dots/inch and a 35-μm spot size.

10.2.2 Landscape lens

The landscape lens[4] is a standard camera lens (or "taking" lens) for an object at infinity and has an aperture stop that can be separate from the lens. The old photographers found that if the lens was oversized compared with the limiting aperture (the stop), they could control the aberrations of the lens by changing the location of the stop, as described in Sec. 10.2.1. In that case, the performance of the lens could be evaluated on the basis of standard scanning specifications: spot size, scan length, and resolution (dots/inch). In the design of a diffractive landscape lens, it is useful to compare it with a refractive lens that can accomplish the same task, imaging a distant scene. Given a refractive triplet and a diffractive singlet, both having a 50-mm focal length, an $f/5.6$ speed, and an 18-deg full field of view, we wish to compare these two lenses.

For this diffractive lens we can use the expressions for the Seidel coefficients for a diffractive lens with the stop in contact that were given in Eq. (10.3). Because the object is at infinity, and if we assume the diffractive surface is on a flat substrate, we can set $B = 0$ and $T = -1$. Again taking the limit of a large refractive index, writing H as $-y\bar{u}$ and ϕ as $1/f$, the Seidel coefficients for a diffractive landscape lens become

spherical $S_\mathrm{I} = y^4/f^3,$ [10.13(a)]

coma $S_\mathrm{II} = -y^3\bar{u}/f^2,$ [10.13(b)]

astigmatism $S_\mathrm{III} = y^2\bar{u}^2/f,$ [10.13(c)]

field curvature $S_\mathrm{IV} = 0,$ and   [10.13(d)]

distortion $S_\mathrm{V} = 0.$ [10.13(e)]

As with the scanning lens, the astigmatism can be eliminated by shifting the stop away from the lens. If t is the distance from the stop location in front of the lens to the lens itself and $\bar{y}$ is the height of the chief ray on the lens, then its height on the lens is $\bar{y} = \bar{u}t$. Using the stop shift equations,[1] it can be shown that the aberrations have been changed to

spherical $S_\mathrm{I}^* = y^4/f^3,$ [10.14(a)]

coma $S_\mathrm{II}^* = -y^3\bar{u}\,(t-f)/f^3,$ [10.14(b)]

astigmatism $S_\mathrm{III}^* = y^2\bar{u}^2(t-f)^2/f^3,$ [10.14(c)]

field curvature $S_\mathrm{IV}^* = 0,$ and   [10.14(d)]

distortion $S_\mathrm{V}^* = \dfrac{y\bar{u}^3 t(3f^2 - 3tf + t^2)}{f^3}.$ [10.14(e)]

An examination of these equations shows that if the stop is placed one focal length in front of the diffractive lens ($t = f$), coma and astigmatism are eliminated and the distortion will be $S_\mathrm{V}^* = y\bar{u}^3$. (This arrangement with the stop one focal length in front of a lens is known as a telecentric design, where all chief rays in the image space are normal to the image plane.) And since the field curvature remains zero during the operation, the image, to a third order, has the same quality across the entire field and is limited only by spherical aberration.

How does this compare with a standard Cooke triplet with the same focal length and aperture ($f/5.6$) with an 18-deg full field of view? Buralli and Morris[4] have shown that a diffractive lens 25 mm in diameter ($f/2$) with a 9-mm aperture 50 mm in front of the lens performs as well as a lens with three components. In the case of distortion at the edge of the field (9 deg off-axis), it is 25% less than the triplet. Because the spherical aberration varies as the fourth power of the aperture (Eq. [10.14(a)]), decreasing the aperture to 6.25 mm ($f/8$) reduces the aberration to the point where the system is diffraction limited over the field.

Because scanning lenses move a laser beam over a patterned or printing surface, their operation is, by virtue of the laser source, monochromatic. For an imager, such as a landscape lens, this is not the case. The variation of the efficiency with wavelength, covered in Sec. 4.3.1, is accompanied by a variation of the Seidel aberrations. This is introduced by way of the power of the lens, which varies with wavelength as

$$\phi(\lambda) = (\lambda/\lambda_c)\phi_c, \qquad (4.36)$$

where λ_c is the central wavelength in the range of wavelengths, discussed in Chapter 4. Here it is the design wavelength. In this telecentric geometry, the only third-order aberration that is affected by polychromatic operation is spherical aberration:

$$S_{\mathrm{I}} = S_{\mathrm{I}}^{*} = y^4/f^3 = y^4/f_c\,(\lambda/\lambda_c)^3\,. \qquad (10.15)$$

Obviously, any broadband source could not be imaged with any decent resolution. The gratinglike structure of the diffractive lens ensures that. However, for narrow bands, such as parts of the infrared region, the performance can surpass a germanium $f/3$ lens from on-axis to 10 deg full field.

10.2.3 Diffractive telescopes

The previous examples consisted of single diffractive lenses. Another example of a purely diffractive system is a telescope for infrared scanning. A telescope is usually considered a visual instrument in that it operates by changing the angular magnification of an object as seen by the eye, but it can also be used to change the angular range of an image scanner. In many infrared systems, infrared imagers and cameras consist of simply an imaging lens and a detector. Some of the detectors are linear arrays, so the image needs to be swept across the linear array with some scanning device such as a scanning mirror. Because the system is particularly useful in the infrared region, it is easier to justify the use of diffractive elements since the fabrication requirements are not as severe as they would be in the visible.

By putting a telescope in front of the imager (Fig. 10.5), the field of view of the camera is reduced by the magnifying power of the telescope, and the resolution on the image plane is consequently magnified by that amount. Because the scanner facet is the limiting aperture of the scan system, it is also the entrance pupil of the scanner-imager combination. One of the principles of optical design is that the most efficient transfer of radiation between two optical systems occurs when the exit pupil of the first system is located at the entrance pupil of the second system.

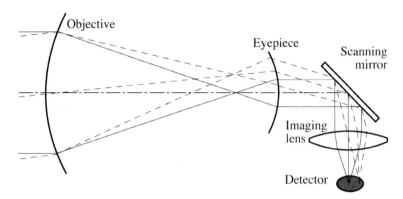

Figure 10.5 Resolution magnification with a diffractive telescope in front of a scanned infrared imager.

A Keplerian telescope design, two positive lenses separated by the sum of their focal lengths, is used because it has an external pupil.[5]

Two elements have six degrees of freedom since there are the substrate curvature for each lens and two phase coefficients (A_1 and A_2) for each lens. The second-order phase coefficients ($\phi = -2\lambda c_1$) are used to establish the powers of the lenses, which in turn determine the angular magnification of the telescope. Thus, there are four degrees to control the third-order aberrations.

The design procedure is similar to that for the diffractive singlets just treated. The aberrations for the object are the same as for the diffractive landscape lens with the stop at the lens and the object at infinity (Eq. [10.13(a)–(e)]). Distortion and field curvature are zero. The eyepiece lens is similar, except that the image is at infinity, so the lens is evaluated in reverse and the stop is shifted to the location of the objective lens. Here the location of the stop is not a variable as it was in the earlier examples. Instead, the substrate curvatures determined by the bending parameters of the two lenses ($B = -2c_s/\phi$) and their fourth-order phase parameters A_2 will be used to correct the third-order performance. The distortion is set to correct angular magnification between the entrance and exiting rays in a manner similar to the scanner lens, but taking into account the magnification of the overall system.

In the case of a Keplerian telescope with a 100-mm objective diameter used at $-5\times$ magnification and operating at a wavelength of 10.6 μm, a compact design was generated with a 200-mm objective focal length ($f/2$) and, by virtue of the $5\times$ magnification, a 40-mm eyepiece focal length. For this system, a clear aperture of 54 mm was needed for the eyepiece to prevent cutting off (vignetting) the input energy. This means the lens must have an f-number of 0.75 or less, a very demanding specification, particularly in fabricating the smallest features (Sec. 4.2.4). One way of dealing with this problem is to split the lens into two lenses of equal power. However, a study[5] showed that little would be gained in optical performance, and since the system was designed for the infrared, the smallest feature size for the two-element design was 16 μm—a size that could be attained. The final design provided wavefront errors of less than 1/60th of a wave at a 4-deg input field angle (20-deg output field angle).

10.2.4 Superzone lenses

In the case of the diffractive lens, we have already determined that the fractures for a kinoform occur at $r_p^2 = 2p\lambda_0 f$ [Eq. (4.13)]. It can be shown that the equation of the profile is expressed as

$$z(r) = \frac{\lambda_0}{n_0 - 1}\left(p - \frac{r^2}{2\lambda_0 f}\right), \qquad (10.16)$$

where $r_{p-1} < r < r_p$ and $n_0 = n(\lambda_0)$. In this instance, the OPD of the kinoform is one wavelength. The feature heights are equal to $\lambda/(n-1)$ and the amount of material to be removed or reshaped is less than in any other designs. However, in some instances, the feature sizes at the edge of an element are so small that they cannot be fabricated.

Although most diffractive optical elements are designed to work in the first diffracted order, it is not required. The feature height can be any integral number of wavelengths (OPD $= q\lambda_0$). This is equivalent to using the grating in a higher order. Such profiles are sometimes called *superzones*. One use of structures with higher depths is to overcome the limitations in fabrication. Because higher-order features have increased feature widths as well as greater depth, the features are great enough to be generated by the available technology.

One strategy is to design a lens with first-order features in the center where the features are coarse and then switch to second and higher orders where the feature sizes become too small to be etched or cut.[6] As noted in Sec. 4.3.1, the efficiency of the diffractive lens falls off from the design wavelength as

$$\eta = \frac{\sin^2\left[\pi\left(\alpha - m\right)\right]}{\left[\pi\left(\alpha - m\right)\right]^2}, \qquad (4.27)$$

where α is the detuning parameter

$$\alpha = \frac{\lambda_0}{\lambda}\frac{n\left(\lambda\right) - 1}{n\left(\lambda_0\right) - 1}, \qquad (4.26)$$

and m is the diffraction order. The plot of efficiency as a function of wavelength for a first-order feature height was shown in Fig. 4.16. However, diffractive lenses can operate at higher orders by designing them with feature heights equal to

$$h = \frac{q\lambda_0}{n(\lambda_0) - 1}, \qquad (10.17)$$

where q is the order for which the lens is designed to operate. However, this high-order diffractive lens will also focus at other orders. It has focal points at

$$f(\lambda) = \frac{q\lambda_0 f_0}{m\lambda}, \qquad (10.18)$$

where λ_0 is the design wavelength, f_0 is the design focal length, and λ is the illuminating wavelength. Note that if the ratio $q\lambda_0/m\lambda$ in Eq. (10.18) is equal to unity, then those wavelengths that satisfy this condition will have a common focus f_0. Thus, depending on the content of the source, the lens can provide reasonable imaging for a number of different wavelengths. The diffraction efficiency for a range of wavelengths about the design wavelength λ_0 is given by an extension of Eq. (4.27):

$$\eta_{q,m} = \frac{\sin^2\left[\pi\left(\alpha q - m\right)\right]}{\left[\pi\left(\alpha q - m\right)\right]^2}. \tag{10.19}$$

In Fig. 10.6 the efficiency of a diffractive lens is plotted as a function of wavelength for three different feature heights. (The dispersion of the material has not been included, so that $\alpha = \lambda_0/\lambda$ for these plots. In a real design, this property must be incorporated to determine actual performance.) Note that in comparison with the efficiency response for a first-order kinoform, the responses for a superzone structure become narrower with higher orders. Also, there are a number of wavelengths where the efficiency is unity. These wavelengths are the same wavelengths with a common focus:

$$\lambda_{\text{foci}} = \frac{q\lambda_0}{m}. \tag{10.20}$$

In some systems where the source consists of emissions from a number of phosphors with narrow bands, the design wavelength and order may be chosen to locate transmission peaks at, or close to, the phosphor peaks. For example, the third-order (center) set of efficiency plots in Fig. 10.6 would be the type of arrangement one might start with in establishing a design.

The quality of image for such a lens is highly dependent on the illumination source. As the source wavelengths depart from the values given in Eq. (10.20) and shown as peaks in Fig. 10.6, the performance is reduced. Faklis and Morris[7] measured the performance of a modest example of such higher-order lenses. They fabricated a lens with $q = 2$ and $\lambda_0 = 640$ nm. The lens also focused the third-order wavelength at 427 nm. They found that the lens focused 91% of the light for these two wavelengths. When used in a telecentric geometry for a landscape lens, described in Sec. 10.2.2, where there is no coma, astigmatism, or Petzval curvature, the lens provides excellent imaging over a wide field of view at selected wavelengths. Others have demonstrated similar operation, but at a considerably higher order, $q = 21$ at $\lambda_0 = 550$ nm. The mold for the profile was fabricated using diamond turning, and an injection-molded lens was cast from it. In monochromatic operation, the lens was not diffraction limited, but was considerably better over a wide number of wavelengths. When imaging a white light source and a natural scene in visible light, this higher-order lens performed as well as a simple refractive lens. The paper contains a discussion of the effects and limitations of broadband operation.[7]

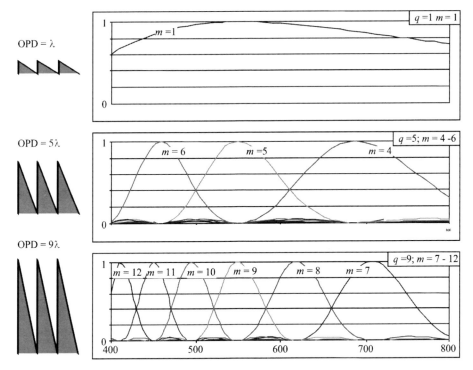

Figure 10.6 Superzone lenses. The diffraction efficiency of the *m*th-diffracted order as a function of wavelength for three lenses with feature heights equal to an integral number of the design wavelength ($\lambda = 550$ nm). Top: First-order design (OPD = λ; compare with Fig. 4.16). Middle: Fifth-order design (OPD = 5λ). Bottom: Ninth-order design (OPD = 9λ).

One of the limitations of these higher-order structures is their off-axis performance. Because of the greater height of the features, they begin to "shadow" the light as light from further out in the field is redirected by these structures. This does not change the image quality as far as aberrations go. Rather, it reduces the off-axis efficiency of the lens below what is calculated using the formulas based on the number of levels and Eq. (10.19).

10.2.5 Staircase lenses

One additional all-diffractive structure is, in a way, the antithesis of the diffractive lenses that have been described in this section. If you consider the shape of the material removed from a classic lens to produce a diffractive lens, you will find that what is left is a lens-shaped profile with flat terraces whose phase differences are equal to 2π (Fig. 10.7).[8] We have described the effect of a diffractive lens on a wavefront that are in the previous parts of this section. What we wish to do here is to look at the effects of a wavefront that are due to the part that was "taken away" from a classic lens to make a diffractive lens.

One of the first consequences of having a structure with surfaces parallel to each other is that a plane wave will be transmitted with no change. That is, a staircase

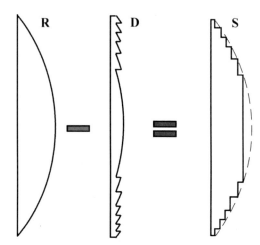

Figure 10.7 Staircase lens. A staircase lens (S) can be described as a refractive lens (R) minus a diffractive lens (D). It consists of a lens-shaped profile with flat terraces.

lens lacks optical power. This might seem to make this a useless element. However, if we look at the off-axis performance, we find that the element introduces field curvature without power. Usually field curvature correction for an imaging lens is achieved by placing a negative lens close to the image plane. There its negative power compensates for the positive field curvature of the imaging lens without contributing to the overall power of the optical system. The difficulty with this solution is that this correction lens must be as large as the image itself. A staircase lens can be located near the imaging lens and need not be any larger than the aperture stop of the lens.

A drawback to these staircase lenses is that besides their field curvature, they are highly chromatic and introduce lateral color (magnification that varies with wavelength), which creates images of different sizes for different wavelengths. Thus, a staircase lens would be most useful in monochromatic applications, unless it should happen to correct for both field curvature and lateral color in the appropriate amounts—an improbable condition.

10.3 Hybrid Lenses

Some examples of hybrid lenses, lenses having both refractive and diffractive power, have already been described in Sec. 4.4. The lenses discussed in this chapter are considerably more complicated, but the basic use of diffractive structures to correct both image and color aberrations remains the same.[9] The addition of diffractive surfaces to an optical system and their variation as part of an optimization can best be understood as part of the exploration of an optical design program that incorporates surface types that can describe a diffractive surface. Failing that, the Sweatt model within a conventional design program can be used. Because all of the examples for this section on hybrid optics are generated through optical design programs, this section provides little information on the basic equations for design and concentrates instead on the results of hybrid designs.

10.3.1 Infrared objectives

An example of an infrared objective is the conventional silicon-germanium doublet operating at $f/2$ with a focal length of 100 mm is shown in Fig. 10.8 along with its ray intercept curves.[10] This lens operates over the middle infrared band (MIR) from 3 to 5 μm. The flatness of the curves indicates that the amounts of spherical aberration and coma are small. The differences in slope between the tangential and sagittal curves are the result of some astigmatism, and the increase in slope with increased field indicates that there is field curvature.

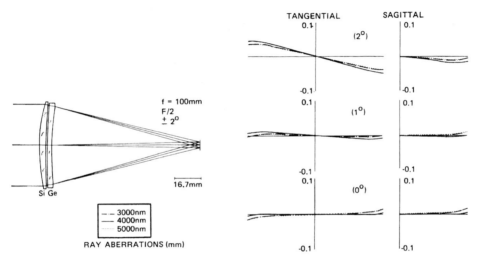

Figure 10.8 Conventional silicon-germanium refractive doublet and its ray-intercept curves.

This conventional doublet can be compared to a single-element germanium meniscus lens with diffractive correction. The fourth- and sixth-order phase terms from the diffractive surface provides color correction and spherical aberration at the center wavelength (4 μm), although the edge wavelengths show spherical aberration (spherochromatism). If an aspheric surface is added to the surface not carrying the diffractive structure, it is possible to achieve performance (shown in Fig. 10.9) that is comparable to the conventional doublet shown in Fig. 10.8.

A Petzval objective is more complicated than the previous imagers. The one shown in Fig. 10.10, along with its performance curves, is used in the MIR over a total field of 1 deg. It is used in front of a two-dimensional sensor array. This example employs four elements and three different materials (Si, Ge, and ZnSe). From the curves it shows itself to be a well-behaved lens. Note that the aberration scale is one-forth of the scale of the previous two lenses.

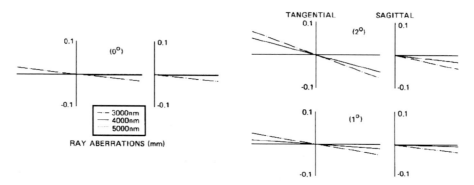

Figure 10.9 Ray intercept curves for a single silicon lens with an asphere and a diffractive surface. The performance is comparable to the doublet shown in Fig. 10.8. There is no spherical aberration or coma, but some astigmatism and field curvature remains, along with secondary color. (Note: the 5000 nm curves do not appear in the original publication.)

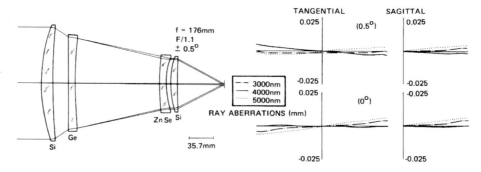

Figure 10.10 Construction and ray intercept curves for an IR Petzval imaging lens with a full field of view of 1 deg.

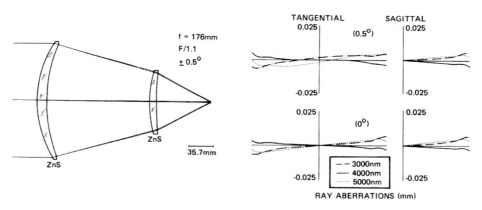

Figure 10.11 A diffractive-asphere version of the Petzval lens with performance comparable to the conventional lens shown in Fig. 10.9.[10]

By using two zinc sulfide meniscus lens, each with an asphere and a diffractive surface on them, it is possible to fabricate a lens of nearly comparable quality (Fig. 10.11). The marginal ray errors are just a bit worse than for the conventional lens, but there are compensations. The weight of the diffractive system is a little over half that of the conventional system (1.1 kg versus 1.9 kg) and the material cost is about one-third that of the first system.

10.3.2 Infrared telescopes

One example of the utility of adding diffractive surfaces is a dual field-of-view, long-wavelength (8 to 12 μm) infrared (LWIR) telescope described by Chen and Anderson.[11] The conventional design, shown in Fig. 10.12(a), consists of six elements, four of which are germanium and two of which are zinc selenide; all surfaces are spherical. When the second, third, and fourth lenses are rotated as a group by 90 deg, removing them from the optical train, the system becomes a narrow field-of-view telescope. The hybrid design [Fig. 10.12(b)] introduces two diffractive surfaces and two aspheric surfaces to the system and at the same time removes the two ZnSe elements. In this case, the second and third elements are rotated out of the way for narrow-field operations.

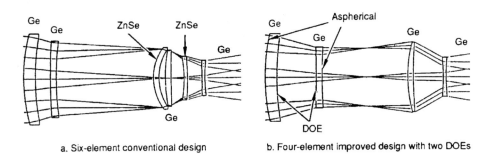

a. Six-element conventional design   b. Four-element improved design with two DOEs

Figure 10.12 Designs for a dual field-of-view LWIR telescope. (a) Conventional design with all spherical surfaces. (b) Hybrid design with two diffractive surfaces and all-germanium elements.

The performance of the improved hybrid design in comparison with the original design is shown in Fig. 10.13. On-axis, the spherical aberration of the hybrid is mainly third order, as opposed to the fifth order for the original design, but the performance is still improved. At the higher field angles, the improvement in performance is quite evident. The reduction in lateral color (variation in magnification as a function of color manifested by the separation between the curves along the *y*-axis) is dramatic. The improvement in performance for the narrow-field configuration is just as impressive. In the review paper that discusses this design,[11] there are six other examples of the use of diffractive optics in infrared systems.

10.3.3 Eyepieces

Another lens system that can be improved by diffractive optics is the eyepiece lens, a high-resolution magnifier. As in the other cases we have looked at, the design of an eyepiece has some special considerations. First, an eyepiece is analyzed in

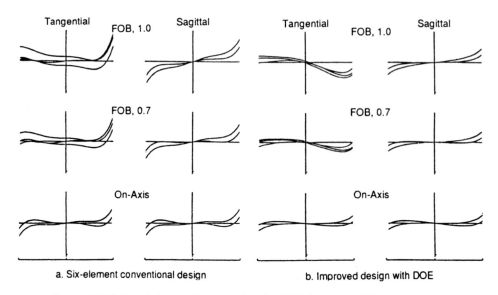

Figure 10.13 Ray intercept curves for dual field-of-view LWIR telescope.

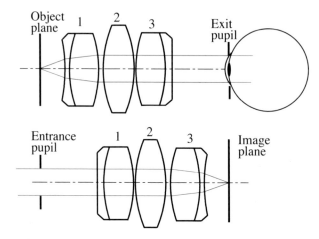

Figure 10.14 Eyepiece geometries. Top: Geometry in which the eyepiece is used. Bottom: Geometry in which rays are traced to determine eyepiece performance. The lens arrangement shown here is used in the classic Erfle eyepiece.

the reverse direction from which it is used. In all our other lenses the object is positioned to the left of the lens and light enters traveling to the right. However, for an eyepiece lens, the object is quite close to the lens and the image to be delivered to the eye is located at infinity (top of Fig. 10.14). It is difficult to trace rays in this manner, so the lens is flipped around front to back and an object is located at infinity and what was the object plane becomes the image plane (bottom of Fig. 10.14). The entrance pupil of the system is located at the pupil of the eye before the geometry was flipped, about 50 mm from the lens. The distance is critical since the eye should be positioned at a comfortable distance from the lens, known as the eye relief. Also, lateral color (the image shows color separation at the edge because magnification varies with color) and distortion should be minimized.

The difficulties of designing an eyepiece lie in its almost impossible specifications. Because we tend to scan a wide field with our unaided eye, we expect to do the same with an eyepiece. When this requirement for a wide field is linked to the need for comfortable eye relief, the sizes of the elements become large. In turn, the size of these elements makes it difficult to correct the Siedel aberrations and hinders color correction. A classic eyepiece, designed and patented by Erfle (Fig. 10.14), is large and bulky, consisting of five lenses. Although it covers a full field of 60 deg, the image at the edge of the field of view shows color despite efforts to correct it.

Because of this difficulty of color correction, the application of diffractive optics, with their extremely large negative dispersion, could be useful in redesigning eyepieces. By eliminating conventional color correction using strong positive crown lenses and strong negative flint lenses, the overall improvement of the monochromatic (Seidel) aberrations should become easier. Also, the short focal lengths of the lenses in an eyepiece contribute to strong field curvature, an aberration completely missing from any diffractive lens, as noted earlier. Diffractive lenses used in place of one or more refractive lenses can help to reduce the overall weight of the optical system, as they did in the Petzval lens example.

Missig and Morris[12] designed two hybrid eyepieces comparable to the Erfle eyepiece shown in Fig. 10.14, which reduced the number of lenses from five to three. In the process they were able to relax the curvatures on the lenses and reduce the weight of the eyepiece by 72%. Starting with the same distribution of lens powers as in the original Erfle design, they replaced the two cemented doublets on each end with plano-convex lenses with diffractive structures on the planar surfaces. This helped to make the fabrication of the diffractive structures easier.

In the first of these designs (Fig. 10.15) two standard glasses, BK7 (for the center lens) and BAK2 (for the end lenses), were used. The designs were optimized in the visible using the C, d, and F lines and the Sweatt model in a standard lens design program. Higher powers of r beyond the quadratic were used in specifying the phase function for the diffractive surfaces. This lens weighs 40% less than the Erfle lens shown in Fig. 10.14, yet it has an eye relief that is 1.36 times larger than the original lens.

In the second of the two designs, BK7 glass is the only material used in conjunction with a single diffractive surface. The initial lens in the first hybrid design is replaced by a meniscus lens. This has the virtue of locating the only diffractive surface in an interior location away from the environment. The weight savings of this second design was even greater than the first. The weight of the second hybrid design was only 30% of the original Erfle and the eye relief was the same as that of the first hybrid design. The aberration correction for both hybrid designs was better than that of the original Erfle, ranging from a modest improvement in distortion to a substantial gain in lateral color (magnification as a function of wavelength). An experimental version of the second hybrid design [Fig. 10.15(b)] was constructed

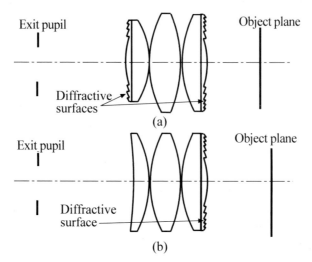

Figure 10.15 Hybrid versions of the Erfle eyepiece. (a) A design with two diffractive surfaces weighs 40% less than the design shown in Fig. 10.14. (b) A design with a single diffractive surface.

and compared with the theoretical calculations. In most instances the measured results were close to the theoretical values. At higher field angles the diffraction efficiency fell off somewhat, owing to an unoptimized blaze profile.

10.4 Thermal Compensation with Diffractive Optics

The optical systems we have demonstrated can be designed by establishing certain tradeoffs that enable the lens to meet the specifications. However, there are other considerations that must be addressed or else the design may be nothing more than an exercise in software use. First, the design has to be capable of being fabricated. For example, a 1-m-diameter diffractive lens may be at the top of the wish list for someone designing an optical satellite surveillance system. It may be easy to establish the phase profile, too, but making such a large element takes a lot of technology and funding. The design considerations that go into fabricating diffractive optics were discussed in Chapters 6 through 9.

Once fabricated, however, the lens may have to fulfill another set of conditions. Once the lens is placed in the real world, it may have to perform under a range of conditions. Some of these, such as dust, humidity, and other environmental variables, may not affect the overall design much. Other variables, such as temperature, could affect the lens performance a great deal. However, there are simple approaches to providing a stable performance for a lens subjected to a range of thermal variation. This section introduces some useful concepts and provides some simple design tools for compensating for thermal changes in the lens environment using diffractive optics.[13-16]

10.4.1 Coefficient of thermal defocus

When the temperature of a lens increases, its focal length f changes as a result of the thermal expansion of the lens and a change in its refractive index. It can be shown that these thermal changes in the focal length are linearly proportional to the focal length itself, so that the changes can be expressed as the amount of change in focal length per unit of focal length. This fractional change in focal length (or optical power ϕ) per unit of temperature change is known as the coefficient of thermal defocus (CTD), x_f:

$$x_f = \frac{1}{f}\frac{df}{dT} = \frac{1}{\phi}\frac{d\phi}{dT}. \tag{10.21}$$

The coefficient for a refractive lens x_f^r consists of three terms: the coefficient of thermal expansion (CTE) α, the change in refractive index of the lens material, and the change in the refractive index of air. (The superscript r on the coefficient indicates the CTD for a refractive lens to distinguish it from a diffractive one.)

$$x_f^r = \alpha - \left(\frac{1}{n_L - n_{air}}\right)\left(\frac{dn}{dT} - n_L\frac{dn_{air}}{dT}\right), \tag{10.22}$$

where n_L and n_{air} are the refractive index of the lens and air, respectively. For example, fused silica has a CTD of -21.1 ppm/°C $= -21.1 \times 10^{-6}$/°C. (The negative sign indicates that the focal length will become smaller as the temperature rises. So an increase in temperature of 10°C will reduce the focal length of a 100-mm-focal length fused-silica lens by 21.1 μm.)

10.4.2 Thermal effects on a mounted lens

If an optical system is to be exposed to a range of temperatures, then some thought must be given to compensating for these temperature changes. Consider the simple example of a lens mounted in an aluminum tube [Fig. 10.16(a)]. Mounted on the end of the tube, one focal length away, is a CCD array. Aluminum has a coefficient of thermal expansion α of $+23.0$ ppm/°C. If the silica lens described earlier is mounted in the tube, the focal point of the lens and the detector surface will separate as the temperature varies from the value at which the system was aligned. Using the case of a 100-mm EFL silica lens, the components will be moved apart by 100 mm $\times$ 10°C $\times$ [23.0 $\times$ 10^{-6} $-$ (-21.1×10^{-6})/°C] $= 44.1$ μm [Fig. 10.16(b)].

To athermalize such a system, the lens or the mount or both must be modified so that the image remains on the CCD. One great simplification is that you do not need to know the focal length of a lens to accomplish this. You have only to match the CTD of the lens to the CTE of the metal tube. For example, a lens of FK52 glass, which has a CTD of 27.5×10^6/°C, would work reasonably well with an aluminum mount, provided the temperature swing was modest. It is possible by combining different metals in some elaborate geometries to match the CTDs for various glasses.

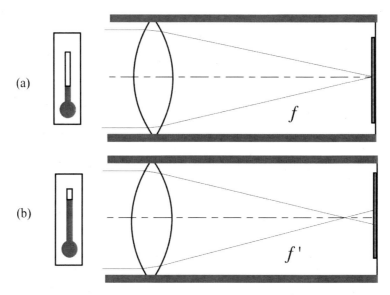

Figure 10.16 Change of focal length with temperature. (a) Lower temperature. Lens is focused on CCD array. (b) Higher temperature. The focal point is no longer at the CCD array because the mounting tube expands and the lens focal length contracts.

10.4.3 Hybrid lens and mount

A diffractive lens also has a CTD, which is given by

$$x_{f,d} = 2\alpha - \frac{1}{n_{\text{air}}} \frac{dn_{\text{air}}}{dT}. \tag{10.23}$$

You can see from the equation that the diffractive lens does not depend on the thermal variation of the refractive index, but that its dependence on the CTE of the glass is doubled. A listing of the CTDs for a number of materials is given in Table 10.1. Because diffractive lenses possesses both optical power and a CTD, the same approach that was used to achromatize a lens can be used to athermalize one.

The basic equations for achromatizing a lens [Eqs. (4.8) and (4.11)] were given in Chapter 4 as

$$\phi = \phi_1 + \phi_2 = (n_1 - 1)c_1 + (n_2 - 1)c_2, \tag{10.24}$$

and

$$\frac{\phi_1}{V_1} = \frac{-\phi_2}{V_2}, \tag{10.25}$$

where the letter subscripts used in Chapter 4 have been replaced by numerical ones. The first of these equations can be rewritten in terms of the fractional powers of

Table 10.1 CTDs for various materials.

Material	$x_{f,r}$	$x_{f,d}$
BK7	0.98	14.20
FK52	27.50	
SF11	−10.61	12.20
Acrylic	315.00	129.00
Polycarbonate	246.00	131.00
Fused silica	−21.10	1.10
ZnS	−36.45	12.91
ZnSe	−28.09	14.51
Ge	−124.95	11.31

the lens, ϕ_1/ϕ and ϕ_2/ϕ as

$$1 = \frac{\phi_1}{\phi} + \frac{\phi_2}{\phi}. \tag{10.26}$$

Equation (10.25) specified the achromatic condition. A more general version of this expresses the V-number for a combination of lenses:

$$\frac{1}{V} = \frac{1}{V_1}\frac{\phi_1}{\phi} + \frac{1}{V_2}\frac{\phi_2}{\phi}. \tag{10.27}$$

Another way of specifying that a lens is achromatized is that the overall V-number must be infinite. When $1/V$ is set equal to zero, we retrieve Eq. (10.25).

A relation similar to Eq. (10.27) can be written for CTDs,

$$x_f = x_1\frac{\phi_1}{\phi} + x_2\frac{\phi_2}{\phi}, \tag{10.28}$$

where the sum of the two terms x_f must equal α, the CTE of the mount, to produce an athermal lens-mount system.

A diffractive surface can be used to compensate for temperature variation in the same way that it was used for color correction in Sec. 4.1.3.1. The difference is that instead of setting the combined CTDs to zero, they are equated to the CTE of the mounting tube. We can use Eqs. (10.26) and (10.28), along with the CTDs of the materials, to set the athermal design. If we stay with our 100-mm EFL fused-silica example, the following values to be entered into the equations are $\phi = 0.01$ mm^{-1}; $x_{f,r} = -21.10$; and $x_{f,d} = 1.10$. Setting x_f equal to the CTE of an aluminum tube ($\alpha = 23$ ppm/°C), we can solve the two equations for the powers of the refractive and diffractive lenses. The result is a negative refractive lens (Fig. 10.17) with a focal length of -101.368 mm ($\phi_1 = -0.009865$) and a diffractive lens of 50.3398 mm ($\phi_2 = +0.019865$) on its plane surface.

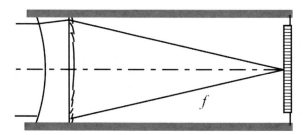

Figure 10.17 Athermal hybrid lens.

As can be seen, a single-material hybrid can be used to correct for either color or temperature. Simultaneous correction of both requires either three materials for a refractive design or two materials for a hybrid design. In such a case, additional terms for a third element need to be added to Eqs. (10.26) to (10.28). These equations are then solved for the powers of the individual lenses. An athermal achromat may be fashioned from a combination of polycarbonate and BK7 glass. The diffractive lens will be formed on the polycarbonate lens. The values to be entered into the equations are $\phi = 0.01$ mm^{-1}; BK7 ($x_{f,r} = 0.98$; $x_{f,d} = 1.10$); poly ($\phi = 0.01$ mm^{-1}; $x_{f,r} = 131.00$; $x_{f,d} = 246.00$; $V_r = 30$; $V_d = -3.45$), where V_r is the refractive V-number of the material and V_d is the diffractive value for all diffractive surfaces, as derived in Sec. 4.4.1. The solution to the simultaneous equations is given in Table 10.2.

Table 10.2 Parameters and powers of an athermal achromat.

Material	Type	V_n	x_n (ppm/°C)	ϕ_n/ϕ	f_n (mm)
BK7	Refractive	64.17	0.98	0.8845	113.058
Polycarbonate	Refractive	30.00	246	0.0609	1642.036
Polycarbonate	Diffractive	−3.45	131.00	0.0546	1831.502
Desired	Hybrid	∞	23.00	1.0000	100.000

References

1. D.C. O'Shea, *Elements of Modern Optical Design*. Wiley, New York, pp. 185 (1985).
2. W.C. Sweatt, *J. Opt. Soc. Am.* **69**, pp. 486 (1979).
3. D.A. Buralli and G.M. Morris, "Design of diffractive singlets for monochromatic imaging," *Appl. Opt.* **30**, pp. 2151–2158 (1991).
4. D.A. Buralli and G.M. Morris, "Design of a wide field diffractive landscape lens," *Appl. Opt.* **28**, pp. 3950–3959 (1989).
5. D.A. Buralli and G.M. Morris, "Design of two- and three-element diffractive Keplerian telescopes," *Appl. Opt.* **31**, pp. 38–43 (1992).
6. J.A. Futhey, M. Beal, and S. Saxe, "Superzone diffractive optics," in Annual Meeting, Vol. 17 of *OSA Technical Digest Series*. Optical Society of America, Washington, DC (1991).

7. D. Faklis and G.M. Morris, "Spectral properties of multiorder diffractive lenses," *Appl. Opt.* **34**, pp. 2462–2468 (1995).

8. J.M. Sasian and R.A. Chipman, "Staircase lens: a binary and diffractive field curvature corrector," *Appl. Opt.* **32**, pp. 60–66 (1993).

9. T. Stone and N. George, "Hybrid diffractive-refractive lenses and achromats," *Appl. Opt.* **27**, pp. 2960–2971 (1988).

10. A.P. Wood, "Using hybrid refractive-diffractive elements in infrared Petzval objectives," *Proc. SPIE* **1354**, pp. 316 (1990).

11. W. Chen and S. Anderson, "Imaging by diffraction: grating design and hardware results," in *Diffractive and Miniaturized Optics*, S.H. Lee, Ed., CR49. SPIE Press, Bellingham, WA, pp. 77–97 (1993).

12. M.D. Missig and G.M. Morris, "Diffractive optics applied to eyepiece design," *Appl. Opt.* **34**, pp. 2452–2461 (1995).

13. G.P. Behrmann and J.P. Bowen, "Influence of temperature on diffractive lens performance," *Appl. Opt.* **32**, pp. 2483–2489 (1993).

14. G.P. Behrmann, J.P. Bowen, and J.N. Mait, "Thermal properties of diffractive optical elements and design of hybrid athermalized lenses," in *Diffractive and Miniaturized Optics*, S.H. Lee, Ed., CR49. SPIE Press, Bellingham, WA, pp. 212–233 (1993).

15. C. Londono, W.T. Plummer, and P.P. Clark, "Athermalization of a single-component lens with diffractive optics," *Appl. Opt.* **31**, pp. 2248–2252 (1992).

16. G.P. Behrmann and J.N. Mait, "Hybrid (refractive/diffractive) optics," in *Micro-optics: Elements, Systems, and Applications*. H.P. Herzig, Ed., Taylor and Francis, London, pp. 259–292 (1997).

Chapter 11

Additional Applications of Diffractive Optical Elements

11.1 Introduction

"Conventional" optics are formed through grinding and polishing techniques. Although they can be made with exquisite quality, the techniques themselves limit the type of the optics that can be produced. As has been noted throughout this text, fabrication methods for diffractive optics are extremely flexible, permitting designers to use a range of optical functions that are not possible with conventional grinding and polishing.

Any discussion of the applications of diffractive optics must be somewhat abbreviated because a complete discussion would require a discussion of almost all the applications of conventional optics, as well as an exposition on a huge range of "nonstandard" optical applications. In Chapter 10, a number of applications of diffractive optics in lens design were discussed. In this chapter we present a survey of additional applications of diffractive lenses as well as gratings, beam shapers, and other types of diffractive optical elements. Just a few of the many applications of these devices include

- Optical computing
- Laser machining
- Imaging systems
- Displays
- Bar code scanning
- Illumination systems
- Position encoders
- Data storage systems
- Photolithography enhancement
- Optical communications
- Biological sensors

This is by no means a complete list of applications. It is intended to provide a sampling of the range of applications of diffractive optical elements. We first present a survey of a variety of diffractive optical elements with example applications. We then discuss in more detail the roles that diffractive optics plays in the area of optical communications and present a detailed example that combines different types of diffractive optics.

11.2 Multiple Lens Applications

In all the examples in Chapter 10, diffractive optics were used to either replace or enhance conventional refractive lenses. The examples in this section use multiple diffractive lenses to perform useful tasks. These configurations are made possible by the flexibility of the fabrication techniques already described for diffractive structures. Until these techniques were developed, the generation of arrays of microlenses (Fig. 11.1) was extremely difficult. Microlens arrays are usually fabricated using the techniques described in Chapters 7 and 8. These techniques can be used to generate both linear and two-dimensional arrays of lenses. Because the lenses can be nested in rectangular and hexagonal arrays, an optical engineer can specify an array of lenses without any wasted area, thereby increasing the efficiency of the system.

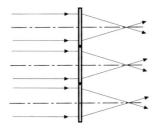

Figure 11.1 Diffractive lens arrays.

11.2.1 Lens arrays for optical coupling

Diffractive optics can be used to capture the light emitted by a laser source for coupling into an optical fiber. The design flexibility provided by diffractive optics allows efficient light collection and coupling, particularly when it is used in conjunction with refractive micro-optics. More applications are making use of arrays of vertical-cavity surface-emitting lasers (VCSELs), in which the individual lasers are fabricated on a wafer with only a small distance (typically from tens to hundreds of microns) between the adjacent lasers. For example, the use of microlens arrays provides a straightforward method for collecting the light from the laser array and coupling it into an array of optical fibers. These arrays can also be used to efficiently couple light from an array of optical fibers or waveguides into another array across a distance, as illustrated in Fig. 11.2.

The type of arrangement shown in the figure can also be used for optical switching with microelectromechanical system (MEMS) mirrors or other switch-

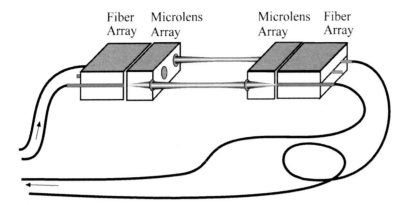

Figure 11.2 Diffractive lens arrays for fiber-fiber coupling.

ing mechanisms.[1,2] Arrays of microlenses are used to collimate the light as it diverges from an array of fibers to strike an optical switching fabric, such as an array of micromirrors. Each of the micromirrors routes the light striking it to a second lens array that collects the light from each directed channel and focuses it back into the output fiber array. Refractive microlenses are also commonly used for this type of application, particularly when the optical switch is intended to operate across a range of wavelengths (from 1310 to 1550 nm for optical communications, for example).

11.2.2 Microlenses for beam steering

The Galilean telescope, a combination of a positive and a negative lens separated by their common focal lengths, is an afocal instrument because the parallel rays entering the positive objective lens are not focused to a point, but instead remain parallel after passing through a negative eyepiece. The power of the telescope is equal to the ratio of the objective's focal length to that of the eyepiece. A trivial case for such a telescope is one with positive and negative lenses of equal magnitude giving a unity magnification. When the two lenses are centered and their optics axes are coincident, the combination acts like slab of glass [Fig. 11.3(a)]. However, if the optic axes of the two lenses are decentered by an amount Δx, a collimated beam traveling parallel to the optic axes of the lenses will be deflected through an angle, $\Delta\theta = \Delta x/f$. To achieve substantial deflections, the focal length of the lenses must be made as short as possible. However, this requires a strong curvature, increasing the overall thickness of the element. Alternatively, a significant amount of relative lens motion Δx is required. These factors can limit the practical use of macroscopic lenses for beam steering.

As an alternative approach, this "fat" lens pair can be replaced by two thin substrates carrying complementary arrays of diffractive lenses, as shown in Fig. 11.3(b). This approach provides the ability to steer beams in much the same manner as just described, but in this case the lenses are nearly planar (and thus can be brought much closer together) and large beam deflections can be achieved with

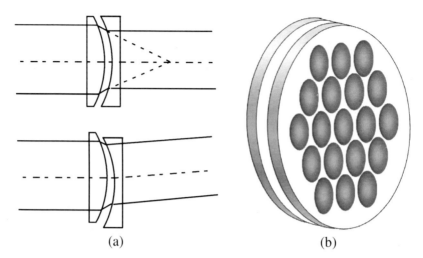

(a)   (b)

Figure 11.3 Beam-steering array. Decentering of a positive-negative pair of lenses causes collimated light to be steered in another direction.

much smaller amounts of lateral motion. A penalty of this approach, however, is that interference between the beams from each of the microlens pairs results in a discretized output "scan" pattern in which the addressable "spots" are separated by an angle $\sim \lambda/d$, where d is the diameter of an individual microlens.[3,4]

The two arrays can be moved using a number of techniques. If large displacements are required, a voice coil can be used, although this approach limits one's ability to rapidly change the beam direction. Piezoelectric devices can provide a more rapid response, with the penalty of smaller displacements.

11.2.3 Lens arrays for sensors

One application of diffractive optics that is particularly useful for sensors with multiple detectors, such as CCD arrays, is the enhancement of their overall detector efficiency by management of the light falling on an array. The ideal detector array would be designed with no dead space between individual detectors, but the demands of fabrication and circuitry design may not permit it. One approach uses diffractive optics to reclaim this dead space by locating an array of microlenses with the same density as the detectors one focal length from the plane of the array (Fig. 11.4). Thus light destined for an inactive region around a detector is collected by the microlens and directed onto the active area of the detector.

Another application for a diffractive optics lens array in conjunction with a detector array is a wavefront sensor. The arrangement, depicted in Fig. 11.5, is similar to the previous geometry: an array of lenses one focal length above the detector array. However, in this case there is a subarray of detectors located beneath each microlens. The biggest difference between this application and the previous one is the way in which the collected energy is processed.

Instead of using the detector to record an image, this geometry is used to measure the shape of a wavefront falling on the array. If collimated, a wavefront nor-

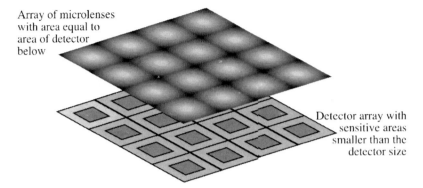

Figure 11.4 Diffractive optics lens array for increasing the efficiency of a detector array.

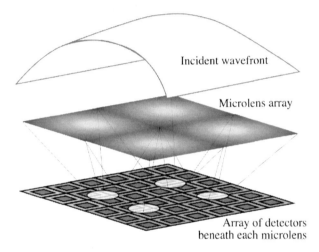

Figure 11.5 Wavefront sensor consisting of an array of detectors beneath each microlens.

mally incident on the lens array would be focused by the lenses to the detector in the subarray on the optical axis of each lens. For a nonplanar wave, light from an area of the wavefront is focused by a lens onto a detector or parts of several detectors, as shown in Fig. 11.5. The relative signals from these detectors can be used to determine the slope of the wavefront at that point on the wave. By collecting the slope information from each of the subarrays beneath each lens and interpolating the values between these points, a complete description of the wavefront shape can be generated. This type of sensor can be used to measure the performance of an optical system with a source as close as on the same laboratory table, or it may determine the wavefront distortion caused by the atmosphere on light from a distant star.

11.2.4 Beam homogenizers

Although it is easy to generate a Gaussian beam output for many lasers, some lasers, notably the noble gas-halide excimer lasers operating in the UV, such as a

KrF or ArF laser, have outputs with large radiance variations across the beam. Most applications, however, require that the laser beam have a smooth profile. Some demand a profile with a "top hat" shape, a nearly constant irradiance across the beam. To be able to achieve these beam profiles, the beam must be "homogenized." Homogenization can be achieved by passing the nonuniform beam through a focusing lens (focal length, f_1) and then through an array of microlenses of focal length f_2, as shown in Fig. 11.6.[5] The different areas of the incident beam pass through the microlens foci and then diverge. The contributions from the microlenses overlap at the focal point of the first lens, producing a smoothed beam. Using some simple geometry it can be shown that the width of the "top hat" is approximately equal to Df_1/f_2, where D is the diameter of a microlens in the array. The width of the beam can be adjusted by changing the focal length of the initial focusing lens.

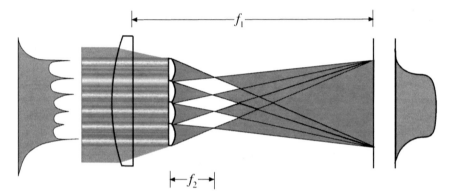

Figure 11.6 Laser beam homogenizer.

11.3 Beam-Shaping Applications

Besides homogenizing nonuniform beams, diffractive optical technology can be used to great advantage by mapping different areas of an input beam of light into differently shaped output beams.[6] Whereas conventional lenses formed by grinding and polishing are limited in this regard (for example, taking a round Gaussian input beam and focusing it to a round, Gaussian spot), diffractive optics can be used to create much more complex beam shapes. In this section, we consider a few simple examples of diffractive beam-shaping elements. A more complex diffractive beam shaper for use in an optical communications system is described in Sec. 11.7.2.

11.3.1 Focusing beam shapers

Most beam-shaping elements can be thought of as a "normal" lens with a twist. Simple designs, optimized to match the incident light beam profile, create a "focused" spot of light in the focal plane of the lens. However, diffractive optics fabrication techniques provide a great deal of flexibility in the shape and irradiance profile of the focal spot that can be created. One of the most straightforward examples of a diffractive beam shaper is a Gaussian to flat-top converter, which performs a 1:1 mapping of a Gaussian input beam into a focal "spot" of uniform intensity, which can be round, square, or any arbitrary shape. The irradiance distribution within the footprint of the focal spot can also be controlled. Similarly, so-called "super-Gaussian" and linear intensity ramps are commonly created. The cross-section and the simulated output of a one-dimensional Gaussian to flat-top beam-shaping element are shown in Fig. 11.7.

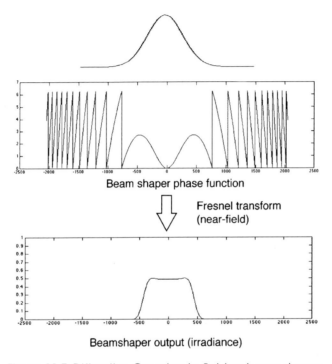

Figure 11.7 Diffractive Gaussian to flat-top beam shaper.

The performance of this device can be understood intuitively by examining the general shape of the diffractive element. The inner portions of the beam shaper direct light from the center of the Gaussian input away from the center in a controlled manner to create a uniform light distribution. Flat-top generators have been used for a wide variety of applications, including laser machining and laser skin treatments in dermatology. Different beam-shaping elements can also be created to map the incident light into much more complicated patterns. One application of this technology is the use of diffractive beam shapers to weld dissimilar materials. Different materials have different thermal properties; diffractive optics can be used to shape the laser beam beyond a simple Gaussian to optimize the "spot" shape for both types of material along the weld joint.[7] A description of optimization for a beam shaper was given in the box at the end of Sec. 5.5. If a second diffractive surface is used, it is possible to shape both the amplitude and the phase of the transmitted wavefront, which allows the creation of useful configurations such as a collimated "flat-top" beam.

11.3.2 Laser resonator design

In addition to the use of diffractive beam shapers outside of a given laser, these powerful and flexible techniques may be applied to almost any laser. We now consider a different case in which diffractive elements are incorporated directly into the laser cavity itself. By choosing the appropriate diffractive design, it is possible to cause the laser to emit light in modes that could not be achieved using conventional techniques. There are several different ways to engineer the chosen lasing mode. In some cases, a diffractive phase plate can be placed between the mirrors of the laser cavity to alter the dominant modes in the cavity. It is also possible to incorporate reflective diffractive structures in place of one or both of the end mirrors of the laser resonator. In addition to achieving results similar to those described in the previous section, for example, a collimated, "flat-top" beam, diffractive structures inside a laser cavity can be used to increase losses in higher-order modes in the laser cavity while enhancing the effective "efficiency" of the laser in the primary lasing mode.[8]

11.4 Grating Applications

Periodic grating structures are perhaps the most fundamental of diffractive optical elements. As discussed extensively in Chapter 5, diffraction gratings are periodic structures that create arrays of light spots (diffraction orders) when illuminated with coherent light. The angular separation of the diffraction orders is determined by the period of the grating and the wavelength of the light illuminating the grating, while the distribution of light between the various diffraction orders is determined by the structure of the grating within a single grating period, or unit cell. With these factors in mind, many different types of patterns can be generated from periodic structures for a variety of applications. We now present several example applications for periodic diffraction gratings.

11.4.1 Beam deflectors, splitters, and samplers

As we noted in earlier chapters, a blazed grating diffracts light selectively into the various orders. When a grating is perfectly "blazed" for a given incident wavelength, all of the diffracted light is directed into the +1 diffraction order. A blazed grating can be used as a simple beam deflector, although there are other, easier ways to deflect a beam of light. When a blazed grating is illuminated with a range of wavelengths, however, the dispersive nature of the grating deflects the different wavelengths of light in the incident beam by different angles, as described by the grating equation. The spectroscopic analysis of the resulting light spectrum can provide a great deal of information about the light source. Because the blaze of a grating can be easily controlled, it efficiently directs light over a band of wavelengths into a single order. Thus, grating spectroscopy was perhaps the "original" application area for diffractive optics. In most of these spectrographs, the diffractive optical elements are used in reflection rather than in transmission.

We have previously shown (Sec. 2.4.1) that a simple binary phase grating with a π phase depth and 50% duty cycle will direct approximately 40.5% of the transmitted light into each of the $+1$ and -1 diffraction orders. Thus, this type of grating can serve well as a simple beamsplitter. By adding more phase levels and changing the design of the grating appropriately, it is possible to engineer the grating to change the ratio of the energy in these two diffraction orders.

A simple but powerful permutation of this concept is a geometry that uses the grating as a beam sampling device. In such a geometry, it is desirable to allow most of the incident light to continue forward, "unaffected," in the zeroth diffraction order while a small amount of the beam is diffracted into a higher order (usually the first order), providing a "sample" of the beam. This can be achieved in a very straightforward manner simply by making the grating much shallower than it would normally be. Presented in another way, we discussed in an earlier chapter how "depth errors," or, more specifically, deviations from the ideal depth, result in more energy in the zeroth order. In the extreme case, a flat piece of glass will have all of the light in the "zeroth order." So a very shallow grating will direct the greatest part of the light in the zeroth order and only a few percent or less in higher orders. By directing the sampled light in the higher order(s) onto a detector, it is possible to monitor, in real time, not only the power levels of a laser beam, but also its profile. The operation of a diffractive beam sampler is illustrated in Fig. 11.8.

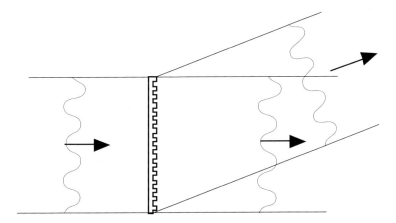

Figure 11.8 Diffractive beam sampler.

11.4.2 Spot array generators

Spot array generators represent a more complex type of diffraction grating than the simple gratings discussed in Sec. 11.4.1. They are usually designed to create an array of equal-intensity diffraction orders, although this is not always the case. The efficiency and uniformity of the spot array output are usually the primary concerns. Although simple designs can be calculated directly either by hand or using a calculator, the vast majority of spot array generators are calculated and optimized using iterative design algorithms, as discussed in detail in Chapter 5.

The larger the array size, the more difficult and time-consuming the calculations. A high-resolution image of an eight-phase level diffractive spot array generator is illustrated in Fig. 11.9(a), and the output from a 9 × 9 spot array generator is shown in Fig. 11.9(b).

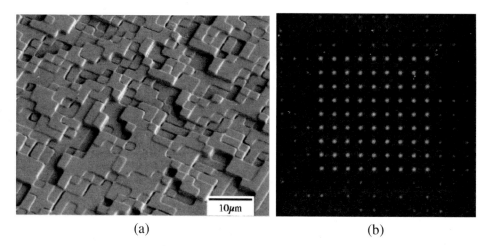

(a)   (b)

Figure 11.9 (a) Eight-phase level diffractive spot array generator. (b) Output from a 9 × 9 diffractive spot array generator. (Courtesy of Digital Optics Corp., Charlotte, NC.)

There is a wide range of applications for spot array generators, including multiple beam splitters, signal distribution for optical interconnects, bar code scanning, and machine vision. As one example, we consider the use of spot array generators for laser machining. Diffractive optics are inherently a good fit because lasers are a required staple of the industry. There are multiple applications in which diffractive optics have been used with great success in this industry. For example, diffractive lenses are regularly used to focus the output of high-power CO_2 lasers to cut sheets of metal and wood. Lower powered lasers are typically used for such applications as laser drilling. For example, lasers are often used to drill holes in printed circuit boards for computer applications. Instead of using a single laser to drill holes in the circuit board one at a time, or many lasers to drill holes in parallel, a spot array generator can be used to create an array of equal-intensity beams from a single laser source to drill a large number of holes simultaneously.

11.4.3 Talbot array illuminators

Another class of diffractive array generator is the Talbot array illuminator. The spot array generators just described form arrays of light spots in the far field of the diffractive optic. In contrast, Talbot array illuminators form light arrays in the near field of the grating. This "nonstandard" application of diffractive optics makes use of the so-called Talbot effect of periodic structures. The Talbot effect refers to the property of periodic structures that creates replicas (sometimes referred to as self-images) of the input field in specific planes in the near field of the grating. These planes occur at integer multiples of the Talbot distance $Z_T = 2\Lambda^2/\lambda$, where Λ is the period of the grating structure and λ is the illuminating wavelength. For example, a binary amplitude diffraction grating illuminated with collimated, coherent light will form a binary amplitude light pattern with the same periodicity as the original grating at integer multiples of the Talbot distance [see Fig. 11.10(a)]. More complicated light patterns are found at fractions of the Talbot distance. Certain phase grating geometries will create arrays of 100% modulated light at fractional Talbot

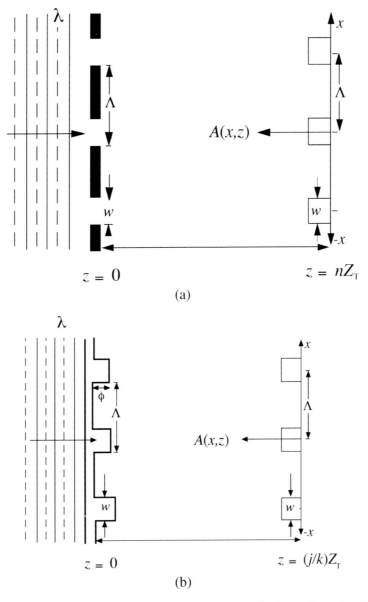

Figure 11.10 (a) Talbot self-imaging from a binary amplitude grating. (b) Operation of a phase-modulated Talbot array illuminator.

distances with essentially 100% efficiency, as shown in Fig. 11.10(b).[9-12] These types of devices have been demonstrated for a wide range of applications, including focal plane enhancement, laser machining,[13] beam steering,[14] and many other uses.[15]

The principles described here work equally well for one- or two-dimensional grating structures. The output from two orthogonally crossed Talbot array illuminators is shown in Fig. 11.11. These types of devices can be used, in appropriate circumstances, as alternatives to the microlens arrays described in Sec. 11.2.1 and the diffractive spot array generators described in Sec. 11.4.2. Unlike the spot array generators, more "spots" can be generated from a Talbot array illuminator simply by illuminating more periods of the grating structure, rather than by redesigning the entire grating.

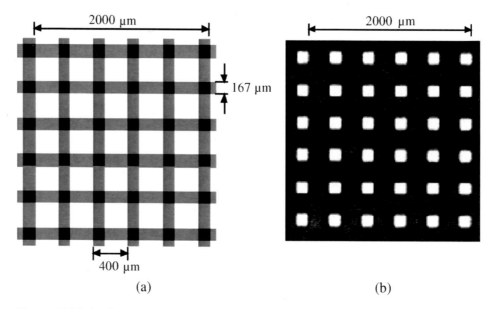

(a)   (b)

Figure 11.11 (a) Schematic illustration of two orthogonally crossed 1D binary phase gratings, each with a period of 400 μm, a duty cycle (w/Λ) of 1/3, and a phase step ϕ of $2\pi/3$. (b) CCD image at one-third of a Talbot distance from these crossed gratings.

11.4.4 Controlled-angle diffusers

Semiconductor manufacturing is an area that has benefited strongly from the use of diffractive optical elements. Lithographic steppers and scanners (discussed briefly in Chapter 7) are the workhorses of the modern semiconductor industry. These tools are capable of lithographically exposing large numbers of photoresist-coated wafers with extreme precision. As the performance of electronic circuits and devices has increased, additional demands have been continually placed on the exposure tools to make ever-smaller features with higher throughputs. As a consequence, any method that allows the fabrication of smaller features and/or increases the depth of focus of the imaging system provides a significant performance benefit for the tool. Controlled-angle diffusers have been shown to provide substantial benefits in this regard.[16] In effect, the field of diffractive optics benefits the technologies that made it possible.

Controlled-angle diffusers are similar in concept and operation to a spot array generator except that the outputs of the diffraction orders are "smeared" together by the diffractive surface to fill in the prescribed area. These devices provide custom, off-axis illumination to get more performance out of existing generations of optical steppers and scanners. The customized illumination patterns serve multiple purposes, including increasing the resolution of the stepper to allow fabrication of smaller feature sizes, and increasing process latitude to increase yield and throughput. These elements can be designed in different ways; some of the most successful approaches require the optimization of literally millions of diffractive orders. Different diffuser patterns optimize different types of features to be made by the stepper (straight lines, lines in orthogonal directions, isolated versus densely packed features, etc.). Circular patterns, annular patterns with different ring diameters, dipoles, and quadrapoles are only a few of the types of diffuser outputs that have been found to increase stepper performance. These devices can also be used to provide customized beam homogenization, as discussed briefly in Sec. 11.2.4. A variety of illumination patterns are shown in Fig. 11.12.

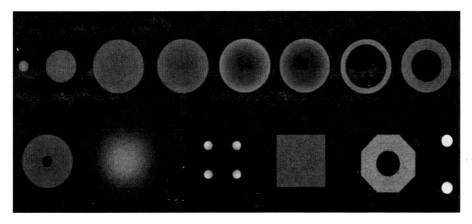

Figure 11.12 Sample outputs from diffractive-controlled angle-diffusing elements. (Courtesy of Digital Optics Corp., Charlotte, NC.)

11.5 Subwavelength Gratings

A particularly interesting set of functions becomes possible when one considers the illumination of gratings with periods that are shorter than the wavelength of the incident light. Assuming normal incidence, it can be shown from the grating equation that only a single transmitted and a single reflected diffraction order are generated by this configuration. The underlying operating principles of these types of gratings are based on effective medium theory, as discussed in Sec. 3.4. The light "sees" a film with an effective refractive index somewhere between the index of the surrounding medium and the index of the substrate material. The effective index can be tailored to an optimum value simply by changing the fill factor of the grating structure. Figure 11.13 shows some sample subwavelength structures and the corresponding index of refraction profiles resulting from effective media theory. It is interesting to consider what types of effects can be generated by changing the grating structures of these subwavelength devices. A wide variety of applications for subwavelength optics have been demonstrated.[17] We now present several examples of the uses of subwavelength optics.

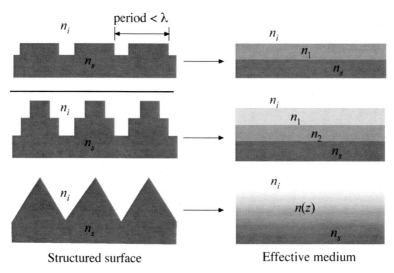

Figure 11.13 Diffractive subwavelength structures and their thin-film equivalents from effective medium theory.

11.5.1 Anti-reflection surfaces and wavelength filters

Reflection occurs whenever the impedance of the medium in which a wave is propagating changes, as in the case of an air/glass interface. In many optical systems, this Fresnel reflection can reduce the efficiency of the system and cause loss of contrast. The simplest way to eliminate Fresnel reflections is to coat one of the interface materials with a third material of appropriate refractive index and thickness. In order to minimize the normal incidence reflection from a boundary between two material regions with indices of n_i and n_s, an index of $n_1 = \sqrt{n_i n_s}$ must be inserted between the two regions with a thickness corresponding to $\lambda/4n_1$. In some instances n_1 exists in the form of a natural material, in which case it can be deposited as a thin film having the required thickness. However, if no such material exists, or if the optical element is to be used in an environment where the thin film is not well suited, then the requisite material can be realized by creating a subwavelength structure. In this case the subwavelength profile is used to create an artificial index of refraction that is used to match the impedance between two adjacent materials. As a result, the use of subwavelength structures for AR surfaces has become more common. It is also possible to use these methods to engineer a multilayer or gradient index stack to obtain AR coatings that operate across wider angles of incidence and wavelength ranges.[18,19]

To design an AR surface using a subwavelength structure, one creates a profile so that the duty cycle and period produce an effective index n_1, according to the relation given earlier, which satisfies the AR condition (as stated earlier). In this case the grating period $\Lambda \leq \lambda/2n_s$ and no material deposition is required, only etching of the host material to the required depth (see Fig. 11.14). The resulting AR surface is as environmentally stable as the material within which it was made. In a similar manner, an AR surface could be constructed using a moth's-eye structure, which was introduced in Chapter 1. In this approach, the index of refraction of the material upon which the incident field is entering is gradually introduced using conically shaped structures, as shown in Fig. 11.14. The goal is to eliminate the impedance mismatch between the two materials by a slow taper so that the incident wave does not see an abrupt junction. Using the analogy presented in Fig. 11.13, the tapered structure can be thought of as a "coating" layer with a gradient refractive index.

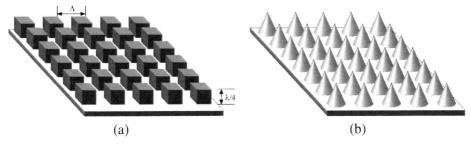

(a)   (b)

Figure 11.14 Illustrations of antireflection surfaces created using subwavelength structures. (a) Two-dimensional binary structure. (b) Moth's-eye structure.

Patterning the subwavelength structures in layers of materials with different refractive indices (for example, patterning gratings in a thin-film dielectric stack on top of a substrate) gives additional degrees of freedom. This approach has been used to enable the fabrication of other devices, such as narrow bandwidth transmission and reflection filters.[20]

11.5.2 Wave plates

In the design of optical systems, it is often desirable to control the orientation, or polarization, of the electric field vector. One example of such an application can be found in the field of telecommunications, where optical modulators are used to control the light entering and leaving an optical fiber. For example, polarization of light that has been scrambled during its propagation down a fiber can affect the signal modulation because the performance of the modulator is polarization dependent. To address this problem, the light exiting the fiber is typically split into two orthogonal polarizations. One of the signals is then rotated to align with the required state of the modulator, after which it can be efficiently modulated by the controlling system. A birefringent material that imparts different phase velocities to different polarizations is required to rotate the polarization states. As a result of differing phase velocities, the two fundamental polarization states of the electric field become out of phase, resulting in a rotation of the polarization direction.

While this procedure is routinely done with naturally birefringent materials such as calcite, it can also be carried out with subwavelength structures. As described earlier, a subwavelength structure can be used to artificially create two different indices of refraction in orthogonal directions by fabricating different grating fill factors in the orthogonal directions. This type of *form birefringence* has been used to produce full-, half-, and quarter-wave plates.[17]

11.5.3 Subwavelength diffraction gratings and lenses

The applications introduced here for subwavelength structures are based on the interaction between the structure and the zeroth-order propagating wave. This means that there are no other propagating orders to consider in either design or analysis. Another set of promising applications for subwavelength structures is based on the interaction between the structure and multiple diffracting orders, namely, diffraction gratings and lenses. Multiple functions have been demonstrated, including blazed gratings, microlenses, and spot array generators.[21–25] In this approach, optical functions are engineered by extreme subwavelength patterning. Binary phase subwavelength optics can have a higher efficiency than diffractive optics with multiple phase levels. As noted in Chapter 2, better approximations to continuous-phase profiles result in improved performance. However, as discussed in earlier chapters, device fabrication becomes much more difficult as the grating period (or zone in the case of a lens) decreases, owing to the increased precision needed to establish alignment between multiple phase levels in the profile. One alternative is to use a two-level device. However, a typical two-level device created using a super-wavelength structure has a maximum efficiency of 41%, according to scalar theory. Thus, in order to obtain high diffraction efficiency using a two-level structure, one must use an extension of the subwavelength design of a linear blazed grating as shown in Fig. 3.12. Although a subwavelength profile may avoid the requirement for precision alignment for multilevel structures, it places stringent requirements on the resolution required in pattern generation and etching. Figure 11.15 shows a fabricated subwavelength lens and its experimental characterization.

Even more challenging types of devices that make use of subwavelength structures have been proposed and demonstrated. For example, photonic bandgap structures[26] (also referred to as photonic crystals) are periodic structures that exhibit a range in wavelengths over which light propagation is forbidden through the structure (thus the name "photonic bandgap"). Example applications for photonic bandgap structures include optical "circuits" that are similar to waveguides,[27] microcavity resonators, and others. Photonic crystal devices offer a great deal of promise for future applications in many areas.

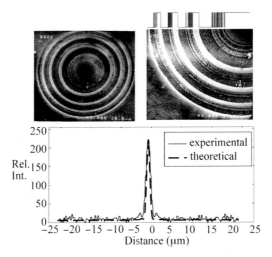

Figure 11.15 (a) Illustration of a subwavelength diffractive lens in fused silica having a diameter of 36 μm, a focal length of 65 μm, and a minimum feature size of 60 nm. The device was designed by Dr. Joseph Mait of the Army Research Laboratory and fabricated by Dr. Axel Scherer of the California Institute of Technology. (b) Overlay of the experimental characterization of the lens and the results predicted from FDTD models of diffraction.

11.6 Integration and Modules

For a decade there has been a movement to integrate active devices (MEMS, laser sources, detectors, etc.) with micro-optics.[28] In the past several years, many applications of fabrication and integration technologies have been demonstrated. This trend toward integration is expected to continue at an accelerated pace. There are a number of factors driving this need for integration of micro-optics into both active and passive devices. Performance, new functionality, miniaturization, and cost reduction are all critical drivers for industries such as optical data storage, sensors, and optical communications. The diffractive (and refractive) micro-optical devices and fabrication methods we have discussed throughout this text are well suited to address these critical needs.

Although both hybrid and monolithic approaches can be used for integration of optical devices, hybrid integration techniques are most common today, and will probably be the practical method of choice for many years to come. The hybrid approach combines "separate" components made by a wide range of processes and materials into optical subassemblies that can then be incorporated into larger-scale optical systems. As the demand for volume manufacturing of integrated photonic systems increases, micro-optical fabrication techniques that are capable of further levels of integration will be in high demand.

Hybrid integration techniques encompass a range of technologies for building optical subassemblies. Prime examples include die bonding of both passive and active components using epoxies and solders, and wire bonding for electrical connections to active devices. By automating these processes, tighter assembly tolerances and higher throughputs than are possible by manual methods can be achieved. As these hybrid integration techniques are automated, it helps to incorporate alignment features into the optical surfaces as part of the fabrication process. These features can be as simple as fiducial patterns that allow visual alignment of one structure relative to another, or patterned metal pads for die bonding or wire bonding. In some cases, it is possible to lithographically fabricate surface relief structures that allow passive alignment and assembly of micro-optics, thereby reducing assembly costs.[29-33]

As the volume of devices increases, additional levels of integration may become desirable. For example, fabrication and integration of optical components at the wafer level allow parallel manufacturing of micro-optical systems.[34,35] By fabricating and bonding arrays of components together at the wafer level and then dicing the wafer into individual subassemblies, economies of scale and cost are achieved. An example of an optical transmitter subassembly fabricated in this manner, shown in Fig. 11.16, is discussed in more detail in the next section.

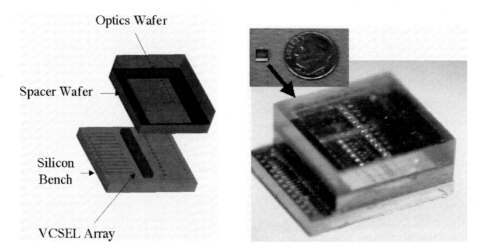

Figure 11.16 Optical transmitter subassembly fabricated from silicon bench, spacer, and micro-optics wafers. (Courtesy of Digital Optics Corp., Charlotte, NC.)

11.7 Example Application Area: Optical Communications

With this overview of different applications for diffractive optical elements, let us now consider in some detail a specific application that greatly benefits from these hybrid devices. Optical communications is a natural area for the application of diffractive optics. This is because laser sources are standard, the miniaturization of optical components and functions is desirable, and a wide range of both conventional and nonconventional optical functions is required for many devices in the industry. By coupling the ability to fabricate large quantities of micro-optics at the wafer level with high precision, diffractive optics will play a major role in the growth of the optical communications industry. This flexibility of diffractive optics can be used in many applications. For example:

- Beam splitters for optical interconnects and signal distribution
- Lenses for coupling light between lasers, fibers, and waveguides
- Gratings for wavelength division multiplexing (WDM) and demultiplexing
- Beam shaping for bandwidth enhancement
- Beam sampling for channel monitors

Refractive micro-optics (discussed briefly in Chapter 8) can also play a major role in optical communications. These devices have very high efficiency and work well across wide wavelength ranges (e.g., 1310 to 1550 nm). Some applications of refractive micro-optics in optical communications include collimating arrays for optical switching, fiber-fiber coupling, laser-fiber coupling for optical transceivers, and light collection from edge-emitting and vertical-cavity surface-emitting lasers. We now consider more specific requirements of micro-optics in optical communications and an example of the application of diffractive optics in communicating with light.

11.7.1 Data communications versus telecommunications

Optical communications systems for data ("datacom") and voice signals ("telecom") are based on the same principles of signal generation, transmission, and detection. However, the specific needs for each type of communication can change the requirements for the optical components and system design. For example, datacom systems typically operate at 850 nm over relatively short distances (up to several hundred meters), where telecommunications applications extend over much longer distances (up to thousands of kilometers) and operate at wavelengths centered at 1310 and 1550 nm.

The objectives for telecom systems, developed in response to long-distance carriers' need for higher signal speeds from city to city, were speed, fidelity, and long life. The resulting systems give excellent performance and reliability, but at a substantial cost. In contrast, datacom systems were developed to meet the needs of businesses and government facilities, where a fast local network was required. The objectives for datacom were low cost and high speed. These datacom systems differ substantially from the telecom format. In time, applications may emerge that are a mix of these two system types as market pressures push them toward common ground, but the distinctions remain important.

11.7.2 Example: parallel hybrid array for data communications

We now consider an example application in data communications that utilizes multiple types of micro-optical components. As noted earlier, optical datacom systems are used to communicate information over relatively short distances. As the need for bandwidth increases, different methods can be used to increase the throughput of datacom systems. One approach is to increase the data modulation speed of a single-channel system. Another is to increase the number of channels in parallel using the lower data modulation rates. One example of such a system consists of

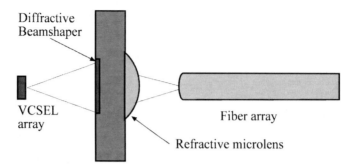

Figure 11.17 Schematic illustration of one channel of an optical transmitter sub-assembly.

a 1 × 12 parallel array of VCSELs with a hybrid 1 × 12 array of diffractive and refractive components to couple light from the VCSEL array into an array of optical fibers.[36] A schematic of one channel of this transmitter module is shown in Fig. 11.17. (Pictures of this module are shown in Fig. 11.16.)

In this system, light from each VCSEL is collected by a diffractive beam-shaping element and directed through a refractive microlens on the other side of the component for additional focusing into an optical fiber. The diffractive beam shaper [shown in Fig. 11.18(a)] in this system performs several different functions and is indicative of the flexibility of diffractive optical elements in optical communications systems. First, the beam shaper contains a diffractive lens that works in conjunction with the refractive lens on the other side to focus the light into the optical fiber. Second, the beam shaper modifies both the amplitude and the phase of light incident from the VCSEL, effectively increasing the amount of information that can be transmitted through the fiber by matching specific transmission modes in the fiber. Finally, the diffractive beam shaper is designed so that the light back-

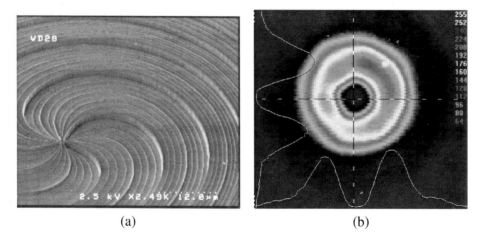

(a)   (b)

Figure 11.18 (a) Diffractive beam shaper for enhanced bandwidth in gradient index fiber. (b) Intensity distribution in focal plane of vortex beam shaper. (Courtesy of Digital Optics Corp., Charlotte, NC.)

reflected from the diffractive surface misses the VCSEL, resulting in a more stable laser operation.

Most datacom applications use multimode gradient index (GRIN) optical fibers. Because of limitations in the manufacturing process for most GRIN fibers, there is a discontinuity in the refractive index profile at the center of the fiber core. Light signals propagating through this discontinuity undergo a phenomenon referred to as differential mode delay that spreads out a signal and limits the rate at which data can be transmitted through the fiber. To minimize this effect, the diffractive element shapes the amplitude and phase of the transmitted light to couple light into the skew modes of the optical fiber so that the light propagating down the GRIN fiber does not cross the centerline defect in the fiber core.[37,38]

11.8 Conclusion

This has been a relatively brief survey of several applications of diffractive optical elements. In addition to these, there are many other applications that have been discussed at various places throughout this text. This wide range of examples demonstrates the enormous flexibility of diffractive optics. Only the laws of physics and the creativity of modern optical designers limit the uses of diffractive optics.

References

1. K. Bergman, N. Bonadeo, I. Brener, and K. Chiang, "Ultra-high capacity MEMS-based optical cross-connects," in *Design, Test, Integration, and Packaging of MEMS/MOEMS 2001*, B. Courtois, J.M. Karam, S.P. Levitan et al., Eds., *Proc. SPIE* **4408**, pp. 2–5 (2001).
2. E. Goldstein, L.-Y. Lin, and J. Walker, "Lightwave micromachines for optical networks," *Opt. Photonics News* (3), pp. 60–64 (2001).
3. W. Goltsos and M. Holz, "Agile beam steering using binary optics microlens arrays," *Opt. Eng.* **29**, pp. 1392–1397 (1990).
4. M.E. Motamedi, A.P. Andrews, W.J. Gunning, and M. Khoshnevisan, "Miniaturized micro-optical scanners," *Opt. Eng.* **33**, pp. 3616–3623 (1994).
5. F. Nickolajeff, S. Hård, and B. Curtis, "Diffractive microlenses replicated in fused silica for excimer laser-beam homogenization," *Appl. Opt.* **36**, pp. 8481–8488 (1997).
6. J.R. Leger, "Laser beam shaping," in *Micro-optics: Elements, Systems, and Applications*, H.P. Herzig, Ed. Taylor and Francis, London, pp. 223–257 (1997).
7. J.B. Hammond, E.G. Johnson, C. Koehler, J. Stack et al., "Diffractive optics for laser welding and bonding," in *Diffractive and Holographic Technologies, Systems, and Spatial Light Modulators VI*, I. Cindrich, S.H. Lee, and R.L. Sutherland, Eds., *Proc. SPIE* **3633**, pp. 206–213 (1999).
8. J.R. Leger, D. Chen, and G. Mowry, "Design and performance of diffractive optics custom laser resonators," *Appl. Opt.* **34**, pp. 2498–2509 (1995).

9. A.W. Lohmann, "An array illuminator based on the Talbot-effect," *Optik* **79**, pp. 41–45 (1988).

10. J.R. Leger and G.J. Swanson, "Efficient array illuminator using binary-optics phase plates at fractional-Talbot planes," *Opt. Lett.* **15**, pp. 288–290 (1990).

11. V. Arrizón and J. Ojeda-Castañeda, "Talbot array illuminators with binary phase gratings," *Opt. Lett.* **18**, pp. 1–3 (1993).

12. T.J. Suleski, "Generation of Lohmann images from binary-phase Talbot array illuminators," *Appl. Opt.* **36**, pp. 4686–4691 (1997).

13. K. Tatah, A. Fukumoto, T. Suleski, and D. O'Shea, "Photoablation and lens damage from fractional Talbot images of Dammann gratings," *Appl. Opt.* **36**, pp. 3577–3580 (1997).

14. M.W. Farn, "Agile beam steering using phased-arraylike binary optics," *Appl. Opt.* **33**, pp. 5151–5158 (1994).

15. K. Patorski, "The self-imaging phenomenon and its applications," in *Progress in Optics*, Vol. 27, E. Wolf, Ed. North Holland, New York, pp. 3–108 (1989).

16. M.D. Himel, R.E. Hutchins, J.C. Colvin, M.K. Poutous et al., "Design and fabrication of customized illumination patterns for low k1 lithography: a diffractive approach," in *Optical Microlithography XIV*, C.J. Progler and A. Yen, Eds., *Proc. SPIE* **4346**, pp. 1436–1442 (2001).

17. J.N. Mait and D.W. Prather (Eds.), *Selected Papers of Subwavelength Diffractive Optics*, MS 166, SPIE Press, Bellingham, WA (2001).

18. D.H. Raguin and G.M. Morris, "Antireflection structured surfaces for the infrared spectral region," *Appl. Opt.* **32**, pp. 1154–1167 (1993).

19. D.H. Raguin and G.M. Morris, "Analysis of antireflection-structured surfaces with continuous one-dimensional surface profiles," *Appl. Opt.* **32**, pp. 2582–2598 (1993).

20. S. Tibuleac and R. Magnusson, "Reflection and transmission guided-mode resonance filters," *J. Opt. Soc. Am. A* **14**, pp. 1617–1626 (1997).

21. W. Stork, N. Streibl, H. Haidner, and P. Kipfer, "Artificial distributed index media fabricated by zero-order gratings," *Opt. Lett.* **16**, pp. 1921–1923 (1991).

22. F.T. Chen and H.G. Craighead, "Diffractive lens fabricated with mostly zeroth-order gratings," *Opt. Lett.* **21**, pp. 177–179 (1996).

23. J.M. Miller, N. de Beaucoudrey, P. Chavel, E. Cambril et al., "Synthesis of a subwavelength-pulse-width spatially modulated array illuminator for 0.633 μm," *Opt. Lett.* **21**, pp. 1399–1401 (1996).

24. S. Astilean, P. Lalanne, P. Chavel, E. Cambril et al., "High-efficiency subwavelength diffractive element patterned in a high-refractive-index material for 633 nm," *Opt. Lett.* **23**, pp. 552–554 (1998).

25. J.N. Mait, A. Scherer, O. Dial, D.W. Prather et al., "Diffractive lens fabricated with binary features less than 60 nm," *Opt. Lett.* **25**, pp. 381–383 (2000).

26. E. Yablonovitch, "Photonic band-gap structures," *J. Opt. Soc. Am. B* **10**, p. 283 (1993).

27. E. Chou, S.Y. Lin, J.R. Wendt, S.J. Johnson et al., "Quantitative analysis of bending efficiency in photonic-crystal waveguide bends at $\lambda = 1.55$ μm wavelengths," *Opt. Lett.* **26**, pp. 286–288 (2001).

28. W.B. Veldkamp, "Overview of micro-optics: past, present, and future," in *Miniature and Micro-Optics: Fabrication and System Applications*, C. Roychoudhuri and W.B. Veldkamp, Eds., *Proc. SPIE* **1544**, pp. 287–299 (1991).

29. W. Singer and K.H. Brenner, "Stacked micro-optical systems," in *Microoptics: Elements, Systems, and Applications*, H.P. Herzig, Ed., Taylor and Francis, London, pp. 199–221 (1997).

30. J. Jahns, "Planar integrated free-space optics," in *Micro-optics: Elements, Systems, and Applications*, H.P. Herzig, Ed., Taylor and Francis, London, pp. 179–198 (1997).

31. M.R. Feldman and Y.C. Lee, Eds., "Micro-optics integration and assemblies," *Proc. SPIE* **3289** (1998).

32. M.R. Feldman, J.G. Grote, and M.K. Hibbs-Brenner, Eds., "Optoelectronic integrated circuits and packaging III," *Proc. SPIE* **3631** (1999).

33. M. Kufner, S. Kufner, P. Chavel, and M. Frank, "Monolithic integration of microlens arrays and fiber holder arrays in poly(methyl methacylate) with fiber self-centering," *Opt. Lett.* **20**, pp. 276–278 (1995).

34. A.D. Kathman, "Integrated micro-optical systems," in *Micro-Optics Integration and Assemblies*, M.R. Feldman and Y.C. Lee, Eds., *Proc. SPIE* **3289**, pp. 13–21 (1998).

35. A.D. Kathman, W. Heyward, and S.W. Farnsworth, "Integrated micro-optical system for LS-120 drive head," in *Gradient Index, Miniature, and Diffractive Optical Systems*, A.D. Kathman, Ed., *Proc. SPIE* **3778**, 104–111 (1999).

36. C.L. Coleman, Y.C. Chen, X. Wang, H. Welch et al., "Diffractive optics in a parallel fiber transmitter module," in *Diffractive Optics and Micro-Optics*, OSA Technical Digest Series, pp. 249–252 (2002).

37. E.G. Johnson, J. Stack, C. Koehler, and T.J. Suleski, "Diffractive vortex lens for mode-matching graded index fiber," in *Diffractive Optics and Micro-Optics*, OSA Technical Digest Series, pp. 205–207 (2000).

38. E.G. Johnson, J. Stack, and C. Koehler, "Light coupling by a vortex lens into graded index fiber," *J. Lightwave. Technol.* **19** (5), pp. 753–758 (2001).

Index